Nutritional Imperialism

REPAST: STUDIES IN THE HISTORY OF NUTRITION

Andrew Ruis, Series Editor

Nutritional Imperialism

How Science Turned Difference into Sickness in China

Hilary A. Smith

JOHNS HOPKINS UNIVERSITY PRESS BALTIMORE

First printed in the United States of America on acid-free paper
9 8 7 6 5 4 3 2 1

Johns Hopkins University Press
2715 North Charles Street
Baltimore, Maryland 21218
www.press.jhu.edu

Library of Congress Cataloging-in-Publication Data is available.

A catalog record for this book is available from the British Library.

ISBN 978-1-4214-5525-9 (paperback)
ISBN 978-1-4214-5526-6 (ebook)

EU GPSR Authorized Representative
LOGOS EUROPE, 9 rue Nicolas Poussin,
17000, La Rochelle, France
E-mail: Contact@logoseurope.eu

Dedicated to the memory of Nathan Sivin (1931–2022),
who saw both science and eating well
as deeply human arts.

When I was studying at university, I felt deeply that to strengthen the country one must first strengthen the citizens, and to strengthen the citizens one must first improve the nutrition in people's diets. I thought exhaustively about how I could use my own studies to contribute to the health of China's masses and wipe away the shame of Westerners falsely labeling Chinese people the "sick men of East Asia."

作者本人在大学学习时期,即深感强国必先强民,强民必先改进人民的膳食营养,极思能以自己所学为我国广大人民的健康有所贡献,以雪洋人诬称中国人是"东亚病夫"之耻。

—Zheng Ji 郑集 (Libin T. Cheng) (1900–2010)
One of the founders of nutrition science in China, reflecting on his career at age 93.

Contents

Figures

Tables

A Note on Conventions

When rendering Chinese words in the Roman alphabet, I use Hanyu Pinyin, the most common system for Romanizing characters. When citing a work that a Chinese person published in English, however, I follow the Romanization and name order that they chose for themselves (for example, Hsien Wu instead of Wu Xian and Libin T. Cheng instead of Zheng Ji). Since these figures wrote sometimes in Chinese and sometimes in English, this may result in multiple renderings of their names in different parts of the book. I have tried to avoid confusing the reader by providing the corresponding name in parentheses, as in "Zheng Ji 郑集 (Libin T. Cheng)." When the source I was looking at used traditional characters, I used traditional characters, and when it used simplified characters, I also used simplified.

For dates, this book uses BC and AD rather than BCE and CE. Changing the words "Before Christ" and "Anno Domini" to "Before Common Era" and "Common Era" without changing the underlying periodization strikes me as a superficial gesture toward inclusivity. Genuine inclusivity would require basing our periodization on something other than the presumed birth year of the central figure in one particular religion. As long as our shared framework continues to derive from that figure, however, we might as well be honest about it.

Finally, recognizing that webpages are more mutable than conventionally published books and articles, I have provided Internet Archive versions of the URLs that I cite in the endnotes. My hope is that these versions will provide readers stable and long-lasting access to the online sources that I have consulted. If they wish to do so, the reader can extract the original URL from the Internet Archive version by deleting everything up to the second https:// in the address.

Acknowledgments

So many people and organizations have helped bring this book into being. I'm both glad to have a chance to recognize them and anxious that I will accidentally leave someone out.

First, there are the colleagues who have been along for this book's whole journey in one way or another. Marta Hanson, Hisa Kuriyama, Michael Nylan and Yi-Li Wu have consistently encouraged me and helped me to better see the big picture. Asaf Goldschmidt, Zhang Ying, Pierce Salguero, Jia-Chen Wendy Fu, Michelle King, Jakob Klein, Angela Leung, Melissa Caldwell, Wayne Soon, and Antje Richter have provided opportunities to present at workshops, edit new scholarship, and publish essays related to the theme of this book. I've learned from thought-provoking conversations with Bridie Andrews, Alice Yao, Sare Aricanli and Joan Judge.

The University of Denver's crack team of teacher-historians, whom I am proud to call my colleagues, have made going to work a pleasure. Carol Helstosky, Susan Schulten, Trishula Patel, Ingrid Tague, Angela Parker, Liz Escobedo, Jonathan Sciarcon, Daniel Melleno, Kim Jones, Bill Philpott, Elizabeth Campbell, Rafael Ioris, and Rhonda Gonzales inspire me with their brilliance and dedication. Our students, with their hard questions and surprising enthusiasms, have shaped this book more than they likely realize.

Many scholars helped me find sources, including Shao-yun Yang, Nathan Hopson, Victor Mair, Wang Gong, Miriam Gross, Minjie Chen, Xiang Li, Jianye He, Cai Shuxuan, and Cheng-Yu Lee. Many thanks, also, to the librarians at Fu Ssu-nien Library and Kuo Ting-yee Library at Academia Sinica, Taiwan, and to the terrific John Moffett, Librarian of the Needham Research Institute's East Asian History of Science Library.

In Cambridge, England I spent what turned out to be the opening months of the COVID-19 pandemic as a visiting scholar affiliated with the Needham Research Institute (NRI). Although we were physically isolated from one another and the libraries and archives were closed, I will always think fondly of the colleagues who weathered that strange time with me. The kindness of the NRI's visiting scholars, who insisted on giving me masks that their families had sent from China when local shops were sold out, still warms my heart. For their humaneness, not to mention their thoughtful contributions to this project, I thank the colleagues I met there, including Mary Brazelton, Jianjun Mei, Christopher Cullen, Geoffrey Lloyd, Catherine Jami, Zhang Meng, Yang Lijuan, Song Yuanming, Yu Jia, Wu Huiyi, Fan Yanni, Bill Mak, and Sue Bennett.

In Taiwan, Shang-Jen Li and Sean Hsiang-lin Lei were instrumental in making my stays as a visiting scholar at Academia Sinica productive, doing everything from providing recommendation and invitation letters to suggesting sources and commenting critically on the project's framing and trajectory. While there, I also had the good fortune to learn from conversations with Chen Hsiu-fen, Pi Kuo-li, Jen-der Lee, Chang Che-chia, Chu Ping-yi, Kevin Chang, Chen Yun-ju, Chen Yu-jen, Johanna Ransmeier, Shen Yubin, SJ Zanolini, JJ Reynolds-Strange, and Lillian Tsay.

Those visiting-scholar stints would have been impossible, and the book might not exist, without the generous funding that institutions have provided. A US Scholar grant from the Chiang Ching-Kuo Foundation and a Jing Brand Scholarship in Chinese Science and Civilisation enabled me to conduct the initial research in Taiwan and Cambridge. Later, a summer stipend from the National Endowment for the Humanities provided time to collect sources for the alcohol chapter and begin to draft it. A Fulbright US Scholar Grant sponsored by the US Department of State and the Foundation for Scholarly Exchange (Fulbright Taiwan) gave a major boost to the project in its final stage. Last and not least, the mini-sabbatical and Professional Research Opportunity for Faculty (PROF) grant funding that the University of Denver provided both improved the quality of the book and hastened its completion. Corinne Lengsfeld, Andrea Stanton, Leasa Weimer, and Noelle Strom all helped me explore and secure grants.

The views, findings, conclusions, or recommendations expressed in this book do not necessarily reflect those of any of the above-named institutions. In fact, I consider it a small miracle that they provide money for scholarly inquiry and relinquish control over the output. I am as grateful for that endorsement of academic independence as for the funding itself.

I wrote many parts of the book in the weekly Write-in-Place sessions that DU faculty hosted. Megan Kelly, Libby Catchings, Rob Gilmor, Matt Hill, Calley Marotta, Alfred Owusu-Ansah, David Riche, Alison Staudinger, and Niki Turnipseed provided many hours of accountability and companionship for drafting and revising. I could always count on the stalwart members of our trans-Pacific writing group, Miranda Brown, Jia-Chen Wendy Fu, Eric Karchmer, and Albert Wu for candid, reflective feedback. The anonymous readers who reviewed the manuscript for Johns Hopkins University Press helped me think about how to reframe the project and highlight important aspects of this history that earlier drafts neglected. At the press, Matt McAdam and Phoebe Oathout's guidance has been invaluable, and Andrew Ruis has provided insightful suggestions at crucial times in his role as series editor. Audra Wolfe of The Outside Reader applied her peerless editorial judgment to a read-through of the entire manuscript when it was close to done.

Even the most expert and punctilious of readers couldn't save me from every oversight, mistake, or wrongheaded argument, so rest assured that they bear no responsibility for the ones you may still find in the pages ahead.

To my family: thank you for being curious travelers. My firstborn served ably as a research assistant and made me laugh at the irony of my worry that his meals have too little protein. Cora was patient when this work took me away from her, without letting me forget that our time together is fleeting and precious. Unflagging cheerleaders Jane and Don Smith have always pitched in figuratively—and, when playing catch with the grandkids, literally. Every kid should have a place to run when the raccoons come. I'm glad my kids do.

A colleague once asked me why I was researching food. Isn't that a bit trivial? Only a person who hasn't faced the relentless work of preparing daily meals could ask such a question. In my family, Babi Hammond has done that work for years, sustaining us with thousands of dinners, not to mention books, movies, music, and laughter. For all that, thank you can never be enough. Love you, shipmate.

Nutritional Imperialism

Introduction

In September 1869, a young American neurologist named George Miller Beard published an article titled "The Diet of Brain-Workers." In it, he suggested that the greatest advancements in human civilization had originated with the best-nourished races and nations. To Beard, writing in the popular magazine *Hours at Home*, this meant the educated English, Germans, French, and particularly his fellow White Americans; their plentiful, diverse, and meaty diets had fueled their creation of laws, military strategy, commerce, science, and literature and thus had made them the "ruling people of the world."[1] By contrast, he asked rhetorically, "What have the natives of South America, the savages of Africa, the stupid Greenlander, the peasantry of Europe, all combined, done for civilization, in comparison with any single beef-eating class of Europe?"[2] It was obvious to him that the diets of the "rice, vegetable and fruit-eating people of ancient and modern times" had not supported the necessary brainpower to make any contribution to human progress.

This self-affirming interpretation of how diet influenced political power and technological progress struck a chord with the editors of many other American publications. Over the course of the subsequent year, an excerpt of Beard's article retitled "The Non-Beef-Eating Nations" appeared in dozens of venues all over the country, from the *Republican Journal* of Belfast, Maine, to the *Presbyter* of Cincinnati, Ohio, to the *Arizona Weekly Journal-Miner* of Prescott, Arizona. The *Saturday Evening Post* published the excerpt, as did the journal of California's Department of Public Instruction.

The shortened, retitled version obscured the main point of Beard's original argument, which had been about class rather than race. Contradicting what scientists studying human metabolism were then suggesting, he had argued that professionals required more and better food than manual laborers. The reprinted excerpts, however, jettisoned Beard's main argument and foregrounded instead his disparagement of "non-beef-eating nations." They show what a keen interest American editors and readers took in the relationship between nutrition, nation, and race.

China required special handling in Beard's analysis because some of his contemporaries admired the Chinese but did not consider them avid meat eaters. This threatened Beard's premise. Thus, he conceded that the Chinese had "developed, it is true, a genius for mechanical arts, and a quiet skill in unique handicraft," but one could not compare such modest distinctions with "those broad purposes of action, that made Rome the mistress of the world, that now compel the eyes of the planet to turn to France, England and America." China, he asserted, had shown no such dynamism. He concluded condescendingly, "The Chinese are as peaceful and inoffensive as we would suppose a nation of rice-eaters might necessarily be."[3]

Do Beard's article and its afterlives tell us anything about science? It is tempting to say that they do not. Although as a neurologist he was an expert on the brain, his piece barely mentioned the organ. His claims about what people of various countries ate rested on anecdote alone. He was pandering to an American sense of superiority in a popular publication, not writing up research results for fellow specialists. All this shows us, you might say, is a typical example of the kind of arrogance that many Americans and Europeans expressed as the age of imperialism neared its height.

Perhaps.

Now, though, let us consider the words of a Chinese nutrition scientist writing almost 60 years later. In 1927, Hsien Wu (Wu Xian 吳憲) published an article titled "Chinese Diet in the Light of Modern Knowledge of Nutrition" in the *Chinese Social and Political Science Review*, an English-language journal whose readership included both Chinese scholars and foreigners interested in China. Wu, sometimes hailed as the "father of Chinese biochemistry," was already one of the country's most prominent figures in the nascent discipline of nutrition science. His article featured much more data than Beard's had included, and its content lay more firmly in Wu's area of expertise. Sprawling across 30 pages, it tabulated the protein, mineral, and vitamin content of different types of foods; estimated the proportion of various grains, legumes, meats, vegetables, and oils in northern Chinese diets; reported the results of feeding experiments on rats; compared the protein and calorie content in Chinese and American diets; and tallied up meat consumption per capita in countries around the world. To a reader today, Wu's work looks far more scientifically rigorous than Beard's.

When it came to the relationship between diet and China's national progress, however, Wu's assessment resembled Beard's to a remarkable degree. The Chinese, he wrote, "are over peaceful, non-persevering, non-progressive, non-enterprising, and are easily contented with the environment in which they find themselves."[4] And like Beard, he suspected that a lack of meat was one cause of this defective national character. He conducted numerous studies comparing rats on vegetable-only diets with rats on mixed meat-and-vegetable diets and extrapolated the results to diagnose the Chinese population. Even if most Chinese people were not showing overt signs of illnesses or syndromes that could be traced directly back to faulty nutrition, Wu thought that inadequate diets over a span of many generations had left his compatriots unable to compete in the modern world. Like Beard's article, Wu's was republished in different forums in both English and Chinese and thus reached a broader audience than the original piece.

Perspectives like the two just described contributed to an ideology that this book calls *nutritional imperialism*. Nutritional imperialism represents the consumption patterns and bodies of people from a wealthy and politically powerful national, racial, or cultural group as a universal standard for healthy diet. It conversely portrays bodies and foodways found among poorer, less-powerful groups as deficient and implies that emulating the eating patterns of the powerful can remedy those deficiencies. Put this way, it is clear that nutritional imperialism is a mindset that one might find in any number of historical settings—any place, in fact, where one social group exercises disproportionate power and influence over another and where people wonder about food's effects on bodies and character. There is no reason why this mindset could only ever privilege the foodways and bodies of people of White European descent or only ever disparage Chinese foodways and bodies. That, however, is the particular variety of nutritional imperialism that this book focuses on.

One reason for focusing on the type of nutritional imperialism that privileges people of Western European descent is that it was in the laboratories of American scientists before and after the turn of the twentieth century that modern nutrition science as we know it today began to emerge.

Those scientists, most of whom were White, perceived the diets and bodies of people like themselves as the norm, and vestiges of that perception remain embedded in some nutrition-science ideas even today. The assumptions that fresh milk is a necessary part of a healthy diet and that wheat is nutritionally superior to rice, for example, appear in popular thinking and official nutritional recommendations all over the world. In other words, of course people of any geographical, national, or racial identity are capable of disparaging the diets and bodies of others; history is replete with examples from ancient times to the present. But it was bourgeois Euro-American diets and White bodies that informed modern nutrition science in its cradle and that continue to exert an outsized influence on dietetics today.

Why focus on China, though, as the foil for nutritional imperialism? After all, the quotation from Beard at the beginning of this introduction suggests that Western hubris about other people's diets was expansive, tarring African, South American, Indigenous American, and Asian foodways with equal relish. Nevertheless, China occupied a special place in the imaginations of early scientists, as Beard's careful, separate treatment of that country demonstrates. In the second half of the nineteenth century, when American scholars were carving the cornerstone of modern nutrition science, immigration was bringing Chinese people and foodways into their consciousness as never before. What Chinese workers ate was a fixation of American labor leaders and politicians, who simultaneously dismissed their (perceived) diets as insufficient and monotonous and feared the men's ability to work tirelessly on such fare.[5] In 1882 the Chinese became the first group ever to be barred from immigration to the United States on the basis of their nationality—a situation that attested to the anxieties attending Chinese identity in the society where early nutrition-science research was taking place.

Simultaneously, Western missionaries and medical men living in China expressed admiration for the healthiness of Chinese diets when compared to those of their own compatriots.[6] Americans supporting alternative views of healthy diet—such as vegetarians—likewise pointed to Chinese foodways as exemplary. Moreover, although by the early twentieth century many Western observers had come to perceive China as the "Sick Man of Asia," they nonetheless recognized China as having had a remarkable past and representing a major, venerable civilization in its own right.

Thus, whether viewed positively or negatively, China was very much on the minds of early nutrition scientists. For Americans, Chinese diets were a complicated, contested topic fraught with political, economic, and civilizational anxiety. When they sought another people, another diet, with which to compare their own, scientists often turned to China and the Chinese as the contrasting case. It is no accident that some of the earliest projects comparing diet across nations and races sought to contrast "Occidental" with "Oriental" metabolism, as this book will explain. In short, to focus on China's place in the development of nutritional imperialism is not arbitrary. China loomed large for the scientists who studied nutrition in the late nineteenth and early twentieth centuries; it should loom large in our histories of nutrition science as well.

The following factors come together to create and sustain nutritional imperialism. First, the imbalance of power and wealth fosters a sense that those who have more are doing something right, while those who have less are doing something wrong. Among the wise choices that those in power have made is eating right. Thus apparent differences in diet become explanations: If the more powerful eat something the less powerful don't, that must contribute to the disparity between them. The key food must have conferred some strength that gave the winners the edge. Maintaining this perspective often requires ignorance of complexity and historical change, because diets are rarely uniform across any large group of people, and eating patterns have changed a great deal over time, even before the modern period. In the version of nutritional imperialism that this book explores, observers saw a number of things as characteristic differences between Chinese diets and those of Europeans and Americans: Where Chinese people ate rice as a staple, Europeans and Americans instead ate wheat, and Chinese people consumed less meat, milk, and alcohol than Europeans or Americans. All of these differences came to be seen as potential explanations for the Chinese nation's weakness.

With that initial belief in place, the second element of nutritional imperialism is the scientific study of the supposed deficit. Experiments abound, many based on questions about how a diet distinctive to a less powerful group has made them sick or weak, or how the bodies of a less powerful group fail to metabolize the esteemed foods efficiently. Results that confirm the expectation of deficiency get more attention than those that refute or complicate it. In the Chinese case, scientific research implied that relying on rice had made people vitamin-deficient, that too little meat had provoked protein deficiency, and that Chinese bodies were lactose intolerant and unable to metabolize alcohol well. In addition, some comparative-metabolism researchers declared that Chinese bodies metabolized all food more slowly than did the bodies of White Americans and Europeans.

Finally, economic incentives help nutritional imperialism persist. These include the incentives of farmers seeking new consumers for their products and governments trying to help their overproducing farmers, those of agribusiness and pharmaceutical companies, and those of researchers who can attract grant money with projects that explore and rectify deficiencies. In Taiwan in the 1950s, American farmers sought (and found, with the help of the US government) consumers to relieve the surplus of milk and wheat they produced. Three decades later, as the dairy industry expanded in mainland China, Chinese companies also became vested in the argument that their fellow citizens need to drink more milk. Pharmaceutical companies inventing remedies for so-called alcohol metabolism deficiency today have an interest in convincing more people that this is a health problem that requires correction. All of these actors have helped to perpetuate the perception that wheat, meat, milk, and alcohol are essential elements of a normal diet and that diets lacking them are defective.

Although all three of these elements converge to create and maintain nutritional imperialism, it is the second, the role of scientists, that particularly interests me. Because of its cultural authority and its claims to produce objective knowledge, science deserves special scrutiny. When we hear a politician or other public figure deride people of a national, racial, or ethnic group or make generalizations about a particular culture, we are bound to approach their words more critically than when we hear a scientist describe a deficiency or disease. An American president calling COVID-19 a "Chinese virus," for example, raises more eyebrows than a textbook description of something called "mitochondrial aldehyde dehydrogenase (ALDH2) deficiency."[7]

This book argues, however, that we ought to subject the latter to the same kind of skepticism that we naturally apply to the former. The textbook entry defines ALDH2 deficiency, colloquially known as "Asian Flush," as "an enzymopathy that impairs ethanol metabolism and affects an estimated 560 million people of East Asian ancestry."[8] Who has determined that having such a constitution—which just means possessing a variant allele of a single gene—is a kind of "deficiency" or "enzymopathy," the -pathy suffix indicating that it is a kind of disorder? How did they come to that conclusion? And given that a gene that slows alcohol metabolism will impact only a person who consumes alcohol, is it accurate to say that this "enzymopathy" affects the lives of every one of the hundreds of millions of people who have the allele? This is only one example. By chronicling the development of the concept of ALDH2 deficiency, along with four other ways in which modern scientists have pathologized Chinese bodies and diets, this book encourages critical and historical thinking about ostensible nutritional disorders said to affect people of one racial or ethno-cultural group more than another.

Scholars I've shared this book with as I've been writing it have sometimes expressed hesitation or even discomfort with applying "imperialism" in this context. For example, when I proposed using "Nutritional Imperialism" as the title of a talk I was giving in China, my host demurred, asking whether it was necessary to use such a sensitive word. We changed the title. Later, in a discussion after a conference paper, a colleague questioned whether "imperialism" was really an apt term for what I had just described. Much of it, he observed, was not about Euro-Americans pathologizing Chinese bodies and practices but about how Chinese nutrition scientists and political leaders had participated in or resisted the development of those ideas—a "two-way street," in his words. And one of my most-respected interlocutors has expressed unease about how some chapters place the words of crude Western supremacists and White scientists with little knowledge of China alongside the work of Chinese nutrition scientists. Doesn't calling that whole complex "imperialism" risk giving the impression that all of these groups were engaged in the same project—namely, the unfair denigration of Chinese bodies, ideas and practices? Doesn't it matter, she asked, where power lies and who is essentializing things Chinese for what purpose?

It does matter, of course. Unlike avowed Western imperialists, Chinese nutrition scientists in the early twentieth century did not demean China's customs or practices to underline the superiority of Europeans and Americans. Instead, they believed their work offered their native country an avenue for self-strengthening. They were brilliant, curious, diligent, and deeply educated people. The same can be said of contemporary scientists investigating genetic variations in metabolism, many of whom are motivated by a desire to make physiological knowledge more racially and ethnically inclusive. But being smart and well intentioned does not preclude a person from inadvertently participating in racist or imperialist ways of thinking.

As the historian Projit Mukharji has pointed out, when most people speak and write about scientific racism, they make it out to be an obsolete "pseudoscience" that we thankfully left behind as we entered the second half of the twentieth century. Imperialism, likewise, is something many people prefer to think of as a module in a high-school history class, cleanly bounded by postwar decolonization and now undisputably a thing of the past. But thinking that way misconstrues the subtlety of both racism and imperialism and obscures ways in which most of us contribute to, or at least acquiesce in, their persistence. Imperialist thinking is "feral," to borrow Mukharji's description of racial thought—a tough, adaptable beast that has survived our attempts to destroy its habitat.[9]

I therefore use the word "imperialism" here as a provocation, an invitation to debate. Chinese scientists did not contribute to this way of thinking in exactly the same way or for the same reasons as their Euro-American counterparts. Late twentieth-century scientists did not talk about race and diet in the same terms as early twentieth-century scientists had. And yet there is more continuity among these geographically and temporally dispersed ideas than most of us are eager to admit. It is precisely because the word raises hackles and stimulates questions that it is useful here. The goal is not to convince the reader that scientists are racists or imperialists. Instead, I wish simply to affirm that scientists live in the same world as everyone else and share the same human tendencies to generalize, to formulate questions based on preconceptions, and to view evidence selectively.

Historians do all these things too, of course. The history of imperialism, diet, and disease in modern China that you are about to read no doubt indulges in some generalizations and omits some important complications. But I have strived to represent the historical scholarship as fairly as I can and to respect the complexity of the primary sources, trusting the reader to decide how convincing they find this account. I hope it will at least encourage you to look critically at assumptions embedded in present-day nutritional knowledge and ask how the political, cultural, and social power disparities of the past have informed them.

A Brief History of Imperialism, Diet, and Nutrition Science in China

Each of the chapters in this book begins in the late nineteenth or early twentieth century. Although their chronological centers differ, each one includes some evidence from across the whole span of modern Chinese history. It will help, therefore, to have some landmarks, a general sense of how imperialism, diet, and nutrition science changed in China over the course of that time. Below, I describe some of the major shifts, dividing the timeline into three periods that historians often use to divide modern Chinese history—namely, the late Qing period (through 1911), the Republican era (1912–1949), and the period after the Chinese civil war (1949–present).

The Late Qing: China's Long Nineteenth Century

The idea that Chinese diets were deficient first emerged during the Qing dynasty in the middle of what in mainland China is referred to as the "century of national humiliation." This period began in 1839 with the onset of the First Opium War with Britain, which ushered in a long interval in which European nations, the United States, and eventually Japan dominated China politically, militarily, and economically. By the turn of the twentieth century, the British, French, Japanese, Americans, and others ran de facto colonies in China's major port cities, the British controlled the Chinese customs service, and many both inside and outside of China believed that the entire empire would soon fragment into foreign-ruled spheres of influence or colonies. At the same time, internal rebellion—including the devastating Taiping Rebellion of 1850–1864, which killed tens of millions of people and devastated the most economically productive region of the empire—destroyed many people's livelihoods. It also further eroded the ability of the Qing government to prevent famine and support those suffering from famines, including the devastating one that ravaged north China from 1876 to 1879. At the same time, many Chinese farmers began to produce cash crops for an international market, which left them vulnerable when prices declined. Thanks to British colonial expansion and industrialization, the average standard of living in the Yangzi Valley region began to lag behind the rapidly growing standard of living in Britain.[10]

This context can help us understand why foreigners' estimation of the value of Chinese diets had declined by the time Beard wrote his essay in 1869. As recently as the early 1800s, Europeans in Asia had admired the dietary customs of the people they encountered there, describing the fresh wholesomeness of the food and the moderation of the eaters as a stark contrast with what they saw as the decadence of their own dietary culture.[11] At that time, many British physicians saw the meat- and alcohol-heavy fare that their countrymen consumed as *less* healthy than what the natives of China ate.[12] As the economic stresses on the Chinese population multiplied and the power imbalance between China and Western nations grew, however, Western disdain for native foodways grew, too. In part this was for the same reason that many late-nineteenth-century Europeans and Americans derided other forms of Indigenous knowledge and practice. They presumed that their military, political, and economic dominance was a natural outcome of what they assumed to be their own superior knowledge and culture—including knowledge about what and how to eat.

Western observers picked up on aspects of Chinese eating patterns that they presumed were representative and traditional—comparatively little consumption of alcohol, for example, virtually no dairy, and a heavy reliance on rice as a staple—and these features became explanations for Chinese subjection to European power. A main problem with these presumptions, as the chapters that follow show, was that they misunderstood or misrepresented both the diversity of Chinese eating patterns and how significantly both Chinese and European diets changed over time.

In any case, in the late nineteenth century perceptions of both Chinese and Western diets were impressionistic and anecdotal at best. Nutrition science was in its infancy. It was represented by chemists and physiologists scattered in European laboratories, such as Justus Liebig, Carl von Voit, and Max Rubner, who had been conducting research on the chemical constituents of food and how food provided energy and replenished damaged or exhausted tissue. By the end of the century American scientists such as Wilbur Atwater had begun to lead the nascent field with systematic surveys of how human bodies used calories.

In China in the nineteenth century, there were no physiology laboratories investigating how bodies metabolized food and no social-science surveys seeking to determine what Chinese people ate. That does not mean that Chinese people lacked dietetic knowledge. As in Europe and the United States at the time, authoritative knowledge about healthy eating still came mostly from physicians drawing on classical guidelines, as well as from folk understandings. Classical Chinese dietetic guidelines, laid out in a rich canon that had begun to take written form a couple of millennia earlier and had continued to expand through the centuries, emphasized eating and drinking in moderation and maintaining a balance among different types of food. The texts finely parsed the qualities of different foodstuffs, assigning them relatively *yin* or *yang* identities as well as flavors (sweet, salty, bitter, sour, and spicy) and thermal natures (very heating, moderately heating, neutral, moderately cooling, and very cooling). The flavors and thermal natures came in sets of five, and that was no accident. In classical Chinese thought the concept of the five phases (*wu xing* 五行)—fire, water, wood, metal, and earth—was an expansive meaning-making matrix that allowed observers to correlate diverse phenomena in order to better understand and intervene in natural processes. Thus, many things came in sets of five—not only flavors and thermal natures but also seasons, stages of life, organs, musical notes, and much else. Knowing that the flavor of a particular foodstuff corresponded to, say, the phase *fire* could help a person understand when, where, and by whom it might most beneficially be eaten, as well as what other foods to pair it with and which to avoid. The classics also included ample knowledge about both how to treat illnesses with food and the sorts of illnesses that food could induce. We will see examples of some of this knowledge in the chapters to come. Folk understandings likewise emphasized that foods were relatively *hot* or *cold* in nature and warned that combining certain kinds of foods, such as crab and persimmon, could be dangerous.

In addition to the classical dietetic tradition and folk beliefs, some modern Chinese eaters turned to Buddhist or Daoist ideas to guide their choices. From a Chinese Buddhist perspective, meat and alcohol were best avoided, as were the "pungent" (*xin* 辛) foods such as onions and garlic, all of which stimulated one's desires and impulses and made it more difficult to follow the sober path to enlightenment. While meat was morally problematic because it violated a prohibition on killing, Daoists sometimes spurned grains and the foods made with them, believing that this helped a person extend their lifespan.

From yin and yang to the five flavors to moral and dietetic prohibitions, this diverse body of knowledge reflected an engagement with the relationships between food, health, self, and environment sustained over the course of many centuries. The nineteenth-century Western observers who perceived China simply as a nation of vegetarians, teetotalers, rice eaters, or dairy-loathers were either uninterested in this knowledge or unaware of its existence.

The Republican Period and the Formation of Nutrition Science in China

After a revolution deposed China's last imperial government in 1911, initial hopes that a new government would stand up to foreign incursions faded quickly in the face of internal disorder and continued external pressures. Foreign powers continued to control concession areas in China's treaty ports. In fact, if anything, imperialist pressure increased as Japan entered the fray and made claims on China equal to or greater than the ones European powers had made. Chinese citizens felt their country's impotence keenly and occasionally reacted against it in fury, boycotting Japanese goods and staging mass protests such as the one on May 4, 1919, that opposed the way imperial powers ignored Chinese sovereignty during the Versailles Treaty negotiations. That protest came to symbolize the larger movement of patriotic youth determined to strengthen their country—thereafter known as the May Fourth Movement—but it made no difference to the way the victorious Allies redistributed formerly Chinese land.

Thus a change in government did not bring about a positive change in China's struggle with imperialism. Nor did it improve diets. In the wake of the imperial government's collapse, warlords vied for power with each other, with an aspiring republican government, and with the Japanese. Farmers continued to struggle, and many workers in the cities subsisted on near-starvation wages. Famines continued to haunt the country, notably across north China in 1920–1921. Unsurprisingly, European and American perceptions of Chinese diets remained mostly negative.

Starting in the 1920s, moreover, Western critiques of China as a vegetarian, milk-avoiding, rice-dependent nation were joined by some new and powerful voices: those of the first generation of Chinese nutrition scientists. To understand them better, let us take a moment to consider how this new discipline emerged in China. Starting in the last years of the Qing empire, many smart and ambitious young Chinese had begun going abroad for education, to Japan, Europe, and North America. This outflow increased from the fall of the Qing government into the 1930s. Most of the students sought higher education in science and technology, with the hope that the skills and knowledge they obtained would help them remedy China's ills and strengthen the nation.

Among the fields that the students chose to pursue was one that was still quite new: nutrition science, a branch of biochemistry. Hsien Wu, the scholar who criticized his countrymen for being "over peaceful" and complacent, was one of the earliest of these students. When he arrived in Boston in 1911 for undergraduate study at the Massachusetts Institute of Technology (to be followed by a PhD at Harvard), he found himself in the same city as the world's first research institute specifically devoted to nutrition science, the Carnegie Institution's Nutrition Laboratory, founded a mere four years earlier. After returning to China in 1920, Wu began work at Peking Union Medical College (PUMC) researching and teaching biochemistry.

In the 1920s Wu and colleagues both Chinese and foreign laid the foundations for nutrition science in China, first establishing departments of biochemistry and physiology in the still relatively new colleges and universities, which had been founded mostly by Western missionaries and philanthropists. These institutions were scattered across China from Beijing to Shanghai and Ji'nan to Chengdu. It was not until the late 1920s that biochemistry and physiology departments began regularly offering nutrition-science courses, but by the 1930s what had begun as a subspecialty was rapidly becoming more visible. Most of the research conducted by the physiology division of the Henry Lester Institute, founded in Shanghai in 1932, focused on nutrition science. In 1934 scientists at Yenching University began publishing *Nutrition Notes*, specifically devoted to reporting new research.

It was the Japanese occupation and the war of resistance starting in 1937, however, that catalyzed the development of an independent discipline of nutrition science in China. Refugees in camps needed to be fed, as did soldiers on the front. Countless people had moved inland to Chongqing and Sichuan from cities in the east and found themselves confronted with a different assortment of foods than they were used to. All of these exigencies made nutrition science seem more urgent and important than it had in the past. At the same time, the country's experts in the subject, formerly scattered among geographically distant metropolises, suddenly found themselves living in much closer proximity to one another in the southwest, having fled eastern cities to escape the Japanese. They did not have the equipment and resources they had been accustomed to at their home institutions, but wartime needs inspired them to tackle new problems collaboratively and establish a firm foundation for Chinese nutrition science in the future.[13] Thus the nation's first national nutrition conference was held in 1941 in Chongqing. The discipline's first Chinese professional association, the Chinese Nutrition Society, was established a few years later, in 1945. The first Chinese scholarly journal devoted to nutrition science began publication in 1947.

The biographies of the Chinese Nutrition Society's first directors and the journal's first editors show that most, like Wu, had earned advanced degrees from American universities. Wan Xin earned his from Purdue (master's degree, 1928), for example, Tang Peisong from Johns Hopkins (PhD, 1930), Libin T. Cheng (Zheng Ji) from Indiana University (PhD, 1934), and Shen Tong from Cornell (PhD, 1939). An early annual report of the Henry Lester Institute, from 1935, lists 22 Chinese researchers, of whom 12 were educated in China at Western-style institutions such as Yenching University, Cheeloo University, PUMC, the Hunan-Yale Medical College, and St. John's University in Shanghai. The 10 who were educated abroad, with the exception of one who studied at the University of Freiburg, all received their degrees from American and Canadian universities: Minnesota, McGill, Indiana, Manitoba, Harvard, Cornell, Johns Hopkins, the University of Pennsylvania, Michigan, and Chicago. Clearly, professional Chinese nutrition science was strongly influenced by developments in America. But the diversity of affiliations—virtually no two of the Lester Institute researchers studied at the same institution—also suggests how diffuse nutritional expertise was at the time, even in the United States. The field was in an early stage of development everywhere.[14]

American nutrition science had, in fact, only recently been transformed, starting in the 1920s with the emergence of the New Nutrition. The expression was drawn from the title of Elmer Verner McCollum's seminal book *The Newer Knowledge of Nutrition*. It referred to a dietetic science focused on micronutrients such as vitamins, rather than on calories and the five broad categories of macronutrients that had earlier been central to scientists' dietary assessments (protein, fat, carbohydrates, minerals, and water). Whereas older nutrition research had taken sufficiency as the goal and had probed the amounts of calories and macronutrients required to keep people alive and functioning normally, the newer nutrition aimed to optimize. What was the *ideal* diet that would not just sustain a person or a nation but enable them to grow and thrive and compete with others? In 1930s and 1940s China, where the upheaval of war and displacement made it difficult for many eaters to achieve even baseline minima, such an idealistic goal made Chinese diets appear even more unsatisfactory. From a New Nutrition perspective, even people who appeared to be adequately fed might be suffering from imperceptible deficiencies that subtly affected their individual state of health and cognition as well as the condition of the whole nation.

In sum, American-inflected chemistry and medicine were the parents of Chinese nutrition science, and war served as midwife. The discipline's genealogy and the circumstances of its birth shaped its character. Together, they produced a body of knowledge primarily concerned with deficits and focused on measuring and maximizing macro- and micronutrients, with reviving the race and improving national competitiveness as one of its central aims. These circumstances may help to explain early Chinese nutrition scientists' tendency to look critically at Chinese eating patterns and to extrapolate from them to nationwide deficiencies.

Other ways of thinking about food and health—classical and folk understandings of the need to balance flavors and thermal qualities, Buddhist and Daoist ideas about fostering equanimity and longevity through food choices—continued to serve as a resource for the many Chinese people who were untouched or even alienated by nutrition science. In addition to these older strands of dietetic knowledge, new dietary practices from all over the globe had taken root in the intellectually freewheeling years following the fall of the Qing government, from those of American Seventh-Day Adventists to those of Japanese intermittent fasters. These, too, persisted, even as nutrition science took institutional shape. Nevertheless, as the Guomindang extended and stabilized its political control over much of China starting in the late 1920s, nutrition-science experts such as Hsien Wu were becoming the standard bearers of officially approved, authoritative knowledge about healthy eating.

Postwar Changes and Continuities

Another seismic change in governance occurred in 1949, when the Chinese Communist Party won the nation's civil war and established the People's Republic of China. As in 1911, Chinese people had high hopes that this change would curtail foreign exploitation of Chinese resources—but unlike the 1911 revolution, for the most part this one did so. There were no treaty ports in the new state and no concession areas. Many foreign-owned businesses were nationalized and foreign missionaries expelled. A large proportion of the Western and Japanese staff of China's universities and research institutes returned to their home countries. The People's Republic followed its own distinctive path until at least the late 1970s, and its leaders have continued to make defending China's sovereignty a chief priority in the decades since.

A couple of qualifications complicate the general picture of Chinese liberation from foreign imperialism in the decades after 1949, however. One is that there were, in fact, two Chinas: While the Communists established their state in Beijing, the remnants of the Guomindang government that had lost the civil war retrenched in Taiwan and maintained a rival regime there. And while the Republic of China did not suffer the same degree of foreign imposition in Taiwan that it had when it was based in mainland China, it relied on American financial and military support. Its relationship with the United States in the 1950s and 1960s, as the following chapters show, influenced and changed the diets of its citizens. The other qualification is that while the former European imperial powers eventually relinquished most of their formal colonial authority in the decades after the Second World War, European states and the United States retained disproportionate influence in the intergovernmental bodies that have structured the postwar world order: the World Bank, the International Monetary Fund, the World Trade Organization, and the United Nations and its constituent organizations. To the extent that those bodies influence China and Taiwan, the United States and former European imperial powers maintain some sway there that is a vestige of their late nineteenth- and early twentieth-century status.

In this same period, diet has improved a great deal overall, in both China and Taiwan. Today's Chinese and Taiwanese citizens have access to more calories and a greater variety of food than they did in 1949, a situation that one imagines Hsien Wu and his colleagues would applaud were they alive today. Again, there are caveats to this general statement. A major one is that, until some years after the death of Mao Zedong in 1976, food was still scarce in mainland China. Urban residents depended on ration cards, and rural ones sometimes had nothing at all to eat, particularly during the mass famine that followed Mao's disastrous collectivization of the countryside in 1958–1961 and killed tens of millions of people. The first decades after the war did not bring dramatic improvement in diets on the mainland, in other words, and sometimes brought dramatic decline instead. It was only once economic reform policies began to show their effects in the 1980s that this situation started rapidly to improve. A second caveat to the general assertion that Chinese and Taiwanese nutritional status has improved is that in both places overnutrition—that is, ill health caused by eating too much food or too much of particular kinds of foods—has in the past two decades become a graver problem than undernutrition, the phenomenon that worried scientists back in the 1930s and 1940s.[15]

Perhaps surprisingly, the disparagement of Chinese eating patterns did not end after China "stood up" against imperialist encroachment in 1949 with the founding of the People's Republic of China. Even as China has asserted its global standing increasingly strongly, and as Western health systems have struggled under the burden of obesity-related disorders, the naturalization of Western-origin diets persists in many quarters along with the idea that Chinese diets and bodies are suboptimal. The concerns that scientists and policymakers had expressed about vitamin and animal-protein deficiencies in the early to mid-twentieth century have been supplanted by concerns about different deficits. Lactose intolerance and alcohol sensitivity have come to be seen as characteristically Chinese physiological challenges. The predominant "deficiency" of concern has changed, but the underlying anxiety that it expresses—that backward habits or embedded cultural defects may be hindering China's return to preeminence—has not. Nor have the benefits to a ruling elite of blaming nutritional problems on eaters rather than on the systems that provide them food. Thus the effects of nutritional imperialism linger, even in the radically different political and social orders of the twenty-first century, and continue to inform things like daily recommended intakes and the content of international food aid. Chinese eaters unable to consume fresh milk or drink alcohol without discomfort are still told that their bodies are abnormal, and researchers endeavor to produce food products and pharmaceutical remedies that will allow them to overcome their incapacities.

In sum, the overall picture of modern change in China and Taiwan in the three areas that concern this book—imperialism, diet, and perceptions of Chinese dietary health—is this: a significant reduction in the influence of foreign powers (with some caveats), a significant increase in the number of calories and the variety of foods available to most people, and continuing concern about supposedly distinctive Chinese deficiencies.

Situating *Nutritional Imperialism* in Historical Scholarship

This study has benefited from the work that other scholars have done to lay the foundation for understanding the relationships among nutrition science, imperialism, and Chinese foodways, building on it in some ways and critiquing it in others. Here, I describe some of the work that has inspired and provoked this research.

Imperialism, Disease, and Diet

Studies of the relationships among imperialism, health, and scientific knowledge are nothing new, but the typical approach of such research has changed over time. The earliest examples echoed the "civilizing mission" conceit of the colonizers themselves. They presented Europeans as imposing order on unruly tropical diseases and environments just as they imposed order on the supposedly uncivilized people native to those environments.[16] By the 1970s and 1980s, historians such as Alfred Crosby and William McNeill were beginning to write in a different vein, exploring how European conquest had dramatically worsened the health of Indigenous peoples in many places. They highlighted in particular the asymmetrical consequences of combining the Eurasian and American "disease pools," as McNeill called them, in the early modern period; the Indigenous people of the Americas suffered far more demographic devastation from these traveling diseases than did Europeans.[17] A third historiographical strand emerged in the 1980s, one that presented medicine, like other technologies, as a "tool of empire" that facilitated European conquest, particularly in areas where vector-borne diseases such as malaria and yellow fever were present.[18]

All of this early work tended to focus on acute infectious disease—smallpox, cholera, influenza—more than on chronic parasitic or nutritional disorders, despite the fact that the latter affected large numbers of people over the course of a lifetime and worsened outcomes when acute infectious diseases swept through. This body of scholarship also tended to focus on what colonizers thought and did, not what Indigenous people thought and did—regardless of whether their authors saw European impact as positive or negative on balance. In the decades since then, however, critical researchers have taken a closer look at how colonizers engineered famine and malnourishment.[19] In addition, more work has emerged that takes Indigenous knowledge, as well as Indigenous assimilation and transformation of Western ideas, seriously.[20]

The study of imperialism's relationship with diet began more recently, over the past fifteen years. Diet played an important role in the cross-cultural encounters that characterized the early modern and modern worlds. Although her book *Cuisine and Empire: Cooking in World History* does not discuss the concept at length, the food historian Rachel Laudan has described most clearly the process by which the diets of the Western elite around the turn of the twentieth century came to be embraced worldwide.

Many intellectuals espoused what Laudan has called "dietary determinism": the notion that you are, profoundly, what you eat. According to this way of thinking, an individual's diet influenced not only their health, as evidenced in height, weight, or luster of hair and skin, but also their intelligence and character.[21] A nation's diet, consumed over the course of decades and centuries, compounded generationally and affected its military and industrial strength as well as its ambition and innovation. In colonies and in politically weaker countries such as China, some leaders viewed their people's disadvantageous position as fundamentally a nutritional crisis requiring rectification. They deliberately adopted Western power cuisine themselves and promoted it among their compatriots, opening the door for Western cuisine to establish itself elsewhere.

Early twentieth-century concern about diets in colonies and weak states was not entirely groundless. Perceived nutritional crises were not only reflections of political anxiety. In many colonial communities and in many communities in China diet *was* insufficient. Ironically, however, given the emphasis on adopting Western diets and learning modern nutrition science, it was often Western power and wealth that had degraded Asian diets by reshaping Asian economies and institutions. The historian Michael Worboys has suggested that chronic malnutrition in colonial India, for example, which British observers "discovered" and represented as a longstanding problem of Indian civilization, was actually a development of the 1920s and 1930s and reflected the way food was then produced, processed, and distributed under British rule.[22] Nor was this phenomenon unique to Asia. Scholars and public-health advocates have shown the same pernicious dynamics at work in colonial and postcolonial Africa.[23] In every case, observers attributed nutrition problems to ignorance of the scientific principles of eating and proposed vitamins, protein supplements, and specific foodstuffs as the simplistic remedies. Western scientific knowledge thus set about remedying a problem that the Western reshaping of economies and foodways had in large part created.

This dynamic is a familiar feature of the relationship between disease and imperialism. Europeans brought vaccination to the Americas, but the pervasive smallpox that they'd seeded centuries earlier was what made vaccination so badly needed there. European administrators brought quinine, a therapeutic agent effective against malaria, to their African colonies from South America. But malaria spread and grew more entrenched as colonial governments and Western companies transformed African agriculture to maximize production, creating canals and reservoirs that bred malaria-carrying mosquitoes and relying on poorly paid and poorly housed workers who were frequently exposed to those mosquitoes. Scholars have helped us perceive this pattern: Imperial conquest spurred social, economic, and environmental changes that caused or exacerbated disease, which authorities then attempted to resolve through narrowly medical means.[24]

In some ways these historical relationships among imperialism, disease, and medicine are, as the saying goes, not even past. For example, the Orientalist assumptions that undergirded imperial powers' sanitary regulations in the nineteenth century, prioritizing diseases that traveled from the colonies to the metropole over those that went the other way, still inform global public-health efforts today.[25] Early twentieth-century beliefs that Native and Black bodies were inherently more vulnerable to tuberculosis than White ones have been echoed—sometimes even cited—in twenty-first-century scientific literature.[26] As for malnutrition, the historian John Nott has shown that "even into the postcolonial period, nutritional intervention was built on undercriticized conclusions about nutrition that were largely formed in the skewed scientific environment of tropical medicine and colonial rule."[27]

This book agrees with Nott, and it suggests that the scientific environment of the past forty years has been skewed, too. Here I draw inspiration from scholars who have examined the social construction of disease since the late twentieth century and shown how we have come to prioritize diagnosing and treating risk pharmaceutically as though risk itself were a disease, as well as how new drugs have contributed to that framing.[28] I argue that just such a redefinition of disease is at work in the invention of lactose intolerance and aldehyde dehydrogenase deficiency. When discussing these conditions I say they were "invented" rather than "discovered" to emphasize that although the underlying chemical and physical mechanisms that researchers discovered are real, framing them as a deficiency requires an act of social imagination. Fixations and blind spots lie in different places than they used to, in other words, but they haven't disappeared—and they have reinforced rather than interrogated old assumptions about nutrition and race.

China, the central focus of this book, has not received as much scholarly attention in the literature on imperialism, disease, and diet as South Asia, Africa, and Southeast Asia. That is understandable—such studies have focused mostly on formal empire. In Asia this has meant French Indochina,

India and Hong Kong under British rule, and Taiwan and Korea when they were governed by Japan.[29] China, which Marxists characterized as "semi-colonial and semi-feudal" in the early twentieth century, never became a formal colony of any one state, but instead suffered the predation of multiple powers. The city of Tianjin, and other treaty ports like it, became what the historian Ruth Rogaski has called a "hyper-colony," carved up into foreign-ruled zones that constituted micro-colonies, while China as a whole remained at least nominally independent from Western governance.[30] This makes it an unusual context in which to observe nutritional imperialism at work, for it was under Chinese governments and in the presence of robust classical and folk knowledge about food that early ideas about dietary deficiencies entered China, and it was in an environment of fierce independence and often strident nationalism that later concepts of deficiency "disorders" developed there.

China beyond Cuisine, Nutrition Science beyond Europe

China-centered food histories have burgeoned over the past 20 years, but they seldom touch on nutrition. They have instead focused mostly on cuisine. There are, for example, studies examining how East Asian cuisines changed over time, how Chinese cuisines spread and were transformed in new settings, how ideas about national and regional Chinese cuisines developed, and how people have defined and redefined "authentic" Chinese cuisine.[31] There are books illuminating the philosophical and ritual contexts of food in early China, but they have little to say about health or medicine.[32] The body of scholarship that addresses both food and health in East Asian history is small, and even in most books of this description, health is addressed in just a few sections or at most a chapter or two. A handful of articles, but very few books, have focused on Chinese ideas about healthy and unhealthy eating.[33]

Meanwhile, a steadily growing body of work on the history of nutrition has given short shrift to the world outside of Europe and North America. Some scholars have explored the development of nutritional paradigms in Europe and others nutritional ideas in the United States.[34] Gyorgy Scrinis has produced a general account of the evolution of nutrition-science ideas in the modern period.[35] Special issues of history journals have focused on the intersection of science, medicine, and food. Some historians have investigated the health effects of particular types of food, such as meat or sugar.

A few have focused in particular on the history of nutrition deficiency disorders.[36]

These histories usefully elucidate how contemporary nutrition science developed, but few examine what was happening outside of Western Europe and North America. In a special issue of the *Journal of the History of Medicine and Allied Sciences*, for example, all five articles—including one on British India—are about European and American perspectives.[37] In an *Osiris* volume devoted to food and science, only three of the seventeen essays focus on a non-Euro-American context.[38] Even when studies in the history of nutrition do venture abroad, it is typically in a colonial context and remains focused on European ideas and policies typically as expressed by European authorities. A 2017 edited volume titled *Setting Nutritional Standards* might more accurately be titled *Europeans and Americans Setting Nutritional Standards*, as its only examinations of a non-Western setting are about French policy in colonial Africa and British policy in colonial India.[39] Nutrition scientists in China shared the concerns of nutrition scientists elsewhere—experts in China worried about protein and vitamin deficiencies just as experts in America did, for example—but few have observed why those concerns took on a special sharpness and assumed distinctive forms in semicolonial and postcolonial settings.

In short, although histories of Euro-American nutrition science and of Chinese cuisine have increased in number and depth, comparatively little research has focused on Chinese nutritional knowledge. This suggests that while scholars and publishers find it easy to recognize China's distinctive contribution to global *cuisine*, they are less ready to perceive China's *science* as anything more than an ersatz version of the kind found in Western countries. There is still an unspoken assumption that what European and American scientists did and thought can stand in for global nutrition science, while what non-Western scientists, healers, eaters, and other dietary experts did and thought was only locally significant.

Nutritional Imperialism contends that China—specifically, Chinese scientists, bodies, and foodways—shaped modern nutrition science, too. Chinese scientists actively contributed to international dietetic research. All too often, Chinese bodies and foodways served as the foil or photo negative against which researchers crafted dietary norms. And while early twentieth-century scientists often ignored or derided classical and folk ideas, in China such ideas coexisted with newer scientific ones, sometimes reinforcing and sometimes undermining them. Older Chinese understandings of food and health constitute even today a resource that we can use to supplement, challenge, and renovate scientific dietetic knowledge.

Accordingly, in each chapter *Nutritional Imperialism* discusses how Chinese eaters and chefs and healers and patients approached the foodstuff at the chapter's center (rice, meat, milk, or alcohol) before the modern period. Because these discussions incorporate textual evidence from very different chronological periods, they may risk giving the impression that before the advent of nutrition science there was one coherent body of Chinese thought and practice that did not change over time—one that, depending on how you read the evidence, was either superior or inferior to the science that would later compete with it. Nothing could be farther from my intention. In fact, the purpose of this evidence is to refute the oversimplified, ahistorical binaries that nineteenth- and early twentieth-century observers posited between Western and Chinese diets by showing how complicated, contextual, and changeable Chinese dietary beliefs and practices have been. Dietary beliefs and practices in Europe and the United States have been equally so. A primary aim of the book is to do for dietetics what other scholars of China have done for medicine: to show how parochial, transient, and locally variable the categories of *both* "Western" and "Chinese" have been.[40]

Outline of the Book

Chapter 1 analyzes comparative racial-metabolism research in the twentieth century and how it helped create a "Chinese normal" distinct from the "standard man" that scientists had previously established by assessing the bodies of White European and American men. Inspired in part by the stereotype that Chinese people were more passive and contemplative compared with active, aggressive Westerners, racial-metabolism researchers tried to determine whether Chinese bodies processed food more slowly than the bodies of White Europeans and Americans.

Each of the subsequent chapters focuses on one example of a way in which researchers believed the Chinese normal fell short of the standard man. Chapter 2 explores the concept of vitamin deficiency, new in the early twentieth century, and how it helped demonize rice-based diets in Chinese societies. When scientists researched a disease called "beriberi" that was erupting among people who ate rice-centered diets, they reinforced an older prejudice that so-called rice eaters were weaker, sicker, and less advanced than those who ate wheat. In chapter 3 we turn to how nutrition scientists fixated for much of the twentieth century on protein and protein deficits. We learn how they came to think of China as a "vegetarian nation" that lagged behind Western countries in part because Chinese people ate too little animal protein.

Chapters 4 and 5 turn to dietary defects invented only in the 1960s and 1970s, though as in the earlier examples, their roots trace back to notions about Chinese civilization popular in the late nineteenth and early twentieth centuries. Chapter 4 examines the concept of lactose intolerance, the pejorative term used to label genetic profiles that cause about three-fourths of the world's population to experience discomfort when drinking milk. Scientists in the 1960s discovered that many Chinese bodies interact with lactose differently than many White bodies. This reinforced an erroneous belief that dairy was something the Chinese had never used or appreciated and became yet another biased indictment of Chinese foodways and bodies. Finally, Chapter 5 explores the history of a physiological state called aldehyde dehydrogenase deficiency, or "Asian Flush." In the 1970s, scientists conducting racially inflected research into alcohol metabolism drew on a belief that Chinese people were inherently moderate drinkers compared with Europeans and Americans. Their work revealed that a substantial percentage of East Asian people have a genetic variation that limits the production of an enzyme important in the digestion of alcohol. In a now-familiar pattern, this difference promptly became a racially coded deficiency, and ever since then, pharmaceutical and supplement manufacturers have been busy inventing remedies for sufferers of so-called Asian Flush.

It is time to provincialize European nutrition science—to do for this form of human knowledge what Dipesh Chakrabarty did for the political thought that educated people around the globe have inherited from the European Enlightenment. As Chakrabarty observed, scholars today cannot avoid working with and through Enlightenment concepts, because they have become an integral part of our identity and activity, regardless of whether we ourselves are of European descent or not. More important, he argued that scholars would not wish to cast aside many of the things that we've inherited; concepts such as secularism, humanism, and individual rights remain our best tools for promoting social justice. But appreciating the value of this body of thought does not mean accepting its every element as true or unquestionable or ignoring the fact that it was created by particular, peculiar groups of people as subject to biases and blind spots as the members of any other group. In Chakrabarty's words, "Provincializing Europe is not a project of rejecting or discarding European thought. . . . European thought is at once both indispensable and inadequate in helping us to think through the experiences of political modernity in non-Western nations, and provincializing Europe becomes the task of exploring how this thought . . . may be renewed from and for the margins."[41]

Likewise, this book recognizes how simultaneously "indispensable and inadequate" modern nutrition science has been. It is surely indispensable. The quantification of diet into ever-finer constituents has enabled researchers to produce supplements and meal replacements that keep fellow humans who might have starved in an earlier age alive—infants who cannot breastfeed, for example, or chronically ill patients who take nourishment through a feeding tube, not to mention Silicon Valley workaholics unwilling to leave their desks to eat. The discovery of vitamin-deficiency disorders and the synthesis of vitamins have helped people avoid and treat disabling symptoms that afflicted many more, particularly the poor and institutionalized, 100 years ago. And like Enlightenment concepts of universal rights, some of the foundational concepts of nutrition science have served and continue to serve justice and equity. For example, research on the food calorie, a new concept in the late nineteenth century, helped to undermine moralistic and subjective judgments about "good" and "bad" diets that had been used to disparage the eating patterns of others. A calorie was a calorie, and some things that the American Anglo-Saxon elite had condemned—such as the one-pot stews of immigrant families—turned out to be abundant in calories as well as the main macronutrients. In some contexts, nutrition-science discoveries also worked against eugenics, suggesting that differences among human groups that some saw as genetic might in fact be nutritional. A change in diet might make a bigger difference than intervening in the gene pool.[42]

On the other hand, nutrition science is also inadequate in ways that we have only just begun to recognize. The reductive fixation on micro- and macronutrients distracts us from more holistic ways of thinking about eating, as Gyorgy Scrinis has suggested.[43] Moreover, there is nothing inherent about scientifically quantifying diets that empowers the disempowered or dismantles racial prejudice. Employers, for example, used the concept of the calorie to undermine demands for wage increases by suggesting that workers would be able to live healthily on their meager pay if only they chose foods with the maximum caloric value instead of more expensive, calorie-light stuff.[44] Much early calorie and physiology research was grounded in the assumption that people of different races had different caloric needs; when the idea of basal metabolism emerged, scientists generally assumed that different races might have faster or slower metabolisms, and they set out to demonstrate that. This had real-world consequences. Finally, while early-twentieth-century nutrition science might seem inoffensive to us compared with early-twentieth-century eugenics, once endorsing eugenics became politically taboo in most places after the Holocaust, nutrition science turned into "one of eugenics' intellectual heirs," as the historian Hannah LeBlanc has observed.[45] One could no longer talk about improving population through selective breeding, but improving it through dietary change remained an acceptable goal.

In other words, nutrition-science research has a complicated legacy. It has served in different contexts to deprive and to uplift. The hope that Chakrabarty expresses—that provincializing Europe might renew thought "from and for the margins" inspires *Nutritional Imperialism*, too. In this case, however, the thing to be renewed is not political theory, but knowledge about eating and sustenance. A purportedly universal science that a priori treats some bodies and practices as inferior not only poorly serves those whose bodies and practices it diminishes but also offers incomplete understanding to everyone. By examining this developing science's premises in a less familiar historical context, we can see their limitations more clearly. Learning about historical resistance to some of nutrition science's more parochial claims can widen our collective understanding of what healthy eating might mean and how variable it can be. Viewing the subject's history from the margins can help renew our collective body of dietetic knowledge and create a nutrition science that is more complete and more flexible.

1

The Standard Man and the Chinese Normal

Comparing Basal Metabolism

Do Chinese bodies process food more slowly than White ones? In the first half of the twentieth century, this question and variations on it motivated a global research program. In the mid-1920s a few experiments hinting that the answer was "yes" intrigued scientists worldwide, including the American biochemist William H. Adolph, who spent most of his career in China. In an article he published in *Scientific American* in 1930, Adolph mentioned a study that suggested Chinese people had a slower metabolism and then speculated about what it might mean. Perhaps, he thought, it meant that Chinese metabolism was "the normal for the human race, and that the rush of the Western civilization has merely produced a higher metabolic rate in the Occident, while the philosophic Orient has avoided this abnormal high tension effect. Or the low metabolism rate for China, if confirmed, may merely indicate that the Chinese can truly live longer on a lower ration than is required by Europeans."[1]

Let us take a closer look at Adolph's musings. As the words "if confirmed" suggest, the state of this research was preliminary. Experiments were scattered and suffered from serious methodological defects, as a few scholars at the time pointed out. But because the idea appealed to stereotypes about the "philosophic" Chinese, studies that seemed to confirm it attracted a lot of attention. Despite contradictory evidence and systematic critiques, the belief that Chinese people had a low metabolic rate clung on tenaciously until at least the 1950s. As for Adolph's concluding words about a lower ration for Chinese people—if they seem ominous, it is because they should. In a modern world relying on technocrats to devise rations for soldiers, students, prisoners, refugees, and more, an ill-founded belief that metabolism varies significantly by race can be downright dangerous.

This chapter explores the ignominious history of racial-metabolism research, and particularly the rise and fall of the hypothesis that Chinese people—like "Oriental" people more generally—digested food more slowly than Americans and Europeans and therefore needed less of it. The studies that racial-metabolism researchers produced were a quintessential example of nutritional imperialism at work. They presented experimental evidence derived from Euro-American bodies as universally applicable to all human beings and experimental evidence derived from non-White bodies as valid only for the particular race that those bodies represented. In other words, what the researchers sought to understand when they analyzed White bodies was the standard human; what they sought to establish when analyzing Chinese bodies was the Chinese normal.

The Calorimeter and the Rise of the Standard Man

Nutrition scientists craved a way to express the fundamental mystery of eating—how food becomes flesh, bones, breath, and heat—in the language of equations. The advent of the food calorie made it possible for the first time to think of diets composed of very different foodstuffs as commensurable. As they came to understand more precisely how bodies burned calories and used nutrients, nutrition scientists also began to estimate how much and what kinds of food human beings needed to consume daily to survive and thrive. Governments and intergovernmental organizations used this information to ration and redistribute food supplies during emergencies and to calculate amounts of humanitarian aid needed by a certain population at a certain time. They initially relied, for such calculations, on an imagined "standard man," an individual in his twenties or thirties weighing 70 kilograms (154 pounds) and leading an active life but one not characterized by hard manual labor—a member of the petite bourgeoisie, perhaps, as were many nutrition-science researchers in the early twentieth century.

The standard man resembled the majority of researchers in another way, too: He was a European or someone of European descent. This was sometimes stated explicitly and sometimes not, but even when a writer did not identify the standard man's race, the data that the writer relied on came from experiments done on the bodies of Europeans and Americans. As early as the 1920s, scientists began to wonder whether the "Caucasian standard," as one put it, could apply equally to people of other races.[2] Did metabolism differ by race? That question absorbed effort and attention well into the 1940s. By the 1980s, the notion of a standard man based on European and American adults had fallen out of favor, but paradoxically, this has not led to a more globally inclusive set of standards. Rather, scientists continue to present standards based on early twentieth-century studies, disproportionately done on White European and American bodies, as universal norms, while devising a host of alternative standards suited to the bodies of people of other races and nations. Below, we will explore the rise, decline, and resurgence of the concept of racial metabolism and in particular its significance for Chinese people.

Around the turn of the twentieth century, studies of metabolism were increasing in scope and precision. The English-language term "metabolism," derived from the Greek word for change, was still relatively new. The French version, *métabolisme*, appears to have been in use by the end of the 1850s and the English word to have come along in the 1870s, but as late as 1897, the author of an article in the *Journal of the American Medical Association* still found it necessary to define the word at the beginning of an article devoted to the topic. He wrote, "In the term metabolism we mean to express the group of phenomena whereby organic beings transform foodstuffs into complex tissue elements, and convert complex substances into simple ones, in the production of energy; in other words the entire process of tissue building, which includes the proper throwing off of unused débris and the elimination of excretory matter, the complete scheme of nutrition, tissue making, production of energy and the elimination of waste matter."[3] In other words, "metabolism" referred to the transformation of one chemical element or one set of compounds into another and the transformation of potential energy into other kinds of energy through the medium of a living body.

Even before the word "metabolism" was coined, scientists had been conducting experiments meant to elucidate how food changed inside bodies. In the 1820s an American army surgeon named William Beaumont performed a series of experiments on an unfortunate fur trapper who, on recovering from being shot, was left with a fistula in his stomach. Through the fistula, Beaumont was able to insert food into his subject's stomach on a string and then pull it out periodically to assess how completely it had been digested. He also removed gastric juice to test its chemical composition and its effects on different kinds of food.[4] In the mid-1800s, the German physiologists Max von Pettenkofer and Carl von Voit devised more precise ways to measure what a body did with food. They collected urine and feces and measured their chemical content, then compared that to the chemical content of the foods the subject had ingested. The main point was to assess how the body processed protein.[5]

It was in the 1890s, when the respiration calorimeter first emerged, that the study of metabolism began to mature. As the instrument's name suggests, its main focus was heat. Scientists had established the "food calorie" or "kilocalorie"—which I will call "calorie" for short—as one of the most essential indices of a food's value.[6] It was the amount of heat required to raise the temperature of one kilogram of water one degree centigrade. Nutrition scientists measured a food's calorie content by burning it in an apparatus called a "bomb calorimeter" and seeing how much that raised the temperature of the water in the apparatus. By using bomb calorimeters to burn foods that most people in America and northern Europe ate, they had developed a knowledge of many foods' calorie content by the turn of the twentieth century.

Knowing the caloric content of foods, alone, was only half of the project of assessing a diet's adequacy. The other half was investigating how many calories a human body needed to fuel its daily activity. For this work, W. O. Atwater and E. B. Rosa built a respiration calorimeter at Wesleyan University in 1894–1895, which would provide a template for similar machines built in subsequent years and decades (Figure 1.1).

Those who volunteered for such experiments, usually young men who were students or scientists, were sealed in a very small room for anywhere from several hours to over a week while the heat (in calories) of all input and output was carefully measured. As one physiologist waggishly described its operation, the respiration calorimeter resembled the bomb calorimeter in that "the subject of the experiment is put in a closed chamber" but differed in that "he is not ignited and burned in oxygen, he is simply allowed to lie comfortably in the insulated chamber, and the amount of heat produced is accurately measured."[7] It was a far more sophisticated way to investigate metabolism than Beaumont's food chunks tied to strings and less invasive for subjects, though not exactly enjoyable. Some experiments in calorimeters less commodious than the Atwater-Rosa kind required the subject to lie in a coffin-like box for several hours with a thermometer inserted 10–12 centimeters into his rectum.[8]

Figure 1.1 Atwater-Rosa respiration calorimeter. The chamber in which the researcher seals the subject is in the center. In the foreground is a refrigerating tank, with a water meter on the right. On the left is an observer's table. W. O. Atwater and E. B. Rosa, "A New Respiration Calorimeter and Experiments on the Conservation of Energy in the Human Body. I." *Physical Review* 9, no. 3 (September 1899): 164.

Atwater took pride in the way that the instrument that he and Rosa had designed could assess total energy exchange holistically. Respiration calorimeters of the Atwater-Rosa type were expensive and complicated machines, however. Not only were they expensive to build—they required an entire room of their own—but also measuring all of the variables and maintaining the machine during experiments required a great deal of instrumentation and at least two skilled operators. That limited their uptake to the most well-resourced institutions of higher learning, even in the United States. In China, where modern-style universities were only just beginning to form, such a machine was an unthinkable luxury. As one observer reported in the *Journal of the American Medical Association* in 1915, "Respiration calorimeters which permit the direct measurement of the heat output are still decidedly uncommon adjuncts to the scientific outfit of the clinic, owing in part to the expense of construction and operation, and in part also to the lack of trained workers who are capable of experimenting in this field."[9] This same author noted, however, that a new technique called "indirect calorimetry" was emerging. It made use of a simpler, more portable apparatus that measured the output of carbon dioxide and the absorption of oxygen in a patient's breath. This provided an acceptable proxy for a subject's resting metabolic rate and made it possible to measure metabolism in more places, not only in the United States and Europe, but also in other parts of the world, including China. To say that it was simpler than the Atwater-Rosa calorimeter, however, does not mean it was simple. The universal respiration apparatus required a rotary blower, electric motor, air moistener, carbon dioxide absorbing apparatus, oxygen cylinder, petroleum manometer, tension equalizer, and balance for weighing absorbers, among other parts.

Despite the rapid spread of instrument-mediated measurement and quantification, nutrition-science research remained quite variable. Different researchers designed their own instruments or modified those of others, and even when they ordered an apparatus of the same design as other researchers, it could differ in ways that affected its function because it was produced by a different manufacturer. Even an identical instrument could produce divergent results in the hands of a technician who had not been trained in the lab where the instrument had originated. Some instruments were so idiosyncratic that only their inventors could coax them into producing data. This made it very difficult to compare results from one country to another or even from one lab to another.[10]

No one expressed more frustration about the incommensurability of nutrition-science data than the American biochemist Francis G. Benedict. Benedict had gotten his start as a chemistry student, earning bachelor's and master's degrees at Harvard and a PhD at the University of Heidelberg. He then joined Atwater's lab at Wesleyan, where he learned the techniques of respiration calorimetry, and in 1907 became the head of the Carnegie Institution's newly founded Nutrition Laboratory, at the time unique in the world in being a research organization focusing on human nutrition. He held the position for the next 30 years, during which time the Carnegie Nutrition Laboratory remained at the vanguard of the discipline.[11] In an effort to combat the variability of instruments and techniques that different scientists used, Benedict spent years visiting his counterparts in laboratories across Europe to directly observe their practices and welcoming European scientists to Boston to gain similar hands-on experience. He developed the Benedict Universal Respiration Apparatus (Figure 1.2) and convinced many scientists to adopt it rather than inventing their own devices.

Figure 1.2 Benedict's universal respiration apparatus. Simpler than the Atwater-Rosa calorimeter, but hardly simple. Notice the nose clip affixed to the reclining subject. Francis G. Benedict, "An Apparatus for Studying the Respiratory Exchange," *American Journal of Physiology* 24, no. 3 (June 1909): 360.

Not only did Benedict want machines and techniques standardized, but he also sought to establish a standard human metabolism. The early studies that Atwater had conducted in his respiration calorimeter had focused on how different kinds of activities might speed up or slow down metabolism. He had compared subjects engaged in brain work (reading or studying) with subjects engaged in physical work (calisthenics or weightlifting), for example.

This yielded data about the different amounts of energy that different sorts of activity required. Surely, though, Benedict surmised, it was more fundamental to measure the underlying metabolism unaffected by the sort of activity a person was engaged in. The measurement he decided on was *basal metabolic rate* (BMR), which some researchers came to call "standard metabolism." BMR was the metabolic rate at complete rest, the energy a body used just keeping itself alive. Subjects prepared for basal metabolism studies by fasting for 12–14 hours and avoiding physical exercise, emotional excitement, and extreme temperatures. During the test they were to remain as still as possible as the machine measured the oxygen and carbon dioxide levels in their breath.[12] The result was a figure supposedly unencumbered by messy variables reflecting environment, activity, or stress. Starting in the early 1920s, Benedict promoted basal metabolism studies as the core activity of the Carnegie Nutrition Laboratory, with the hope that such work would become central to nutrition science worldwide.[13]

A focus on BMR shifted researchers' gaze away from comparing different kinds of work or activity and toward comparing different kinds of people. It quickly became apparent that basal metabolic rate was not uniform across the human lifespan. It was higher in young people and decreased as a person aged. It seemed to be lower in women than in men, and even among young adult men, measured BMR could vary significantly from person to person. To arrive at a standard metabolism, then, required establishing a standard person.

The standard man was already a familiar concept in dietetic literature. It had arisen in the mid-nineteenth century and accompanied the development of statistical thinking and methods.[14] Baron Lyon Playfair, in his 1865 treatise *On the Food of Man in Relation to His Useful Work*, had used the term "standard man" repeatedly, defining it as a "man of healthy digestion" who weighed 150 pounds.[15] He offered a table itemizing the quantities of different kinds of digestive juice secreted by a standard man in 24 hours, including saliva, gastric juices, pancreatic juice, and intestinal juice.[16] What caused Playfair to deem this specimen of a man "standard" is unclear. It seems to have been a person he himself perceived as "normal"—not notably fat or thin, not notably active or sedentary, not diseased, not exceptional in any way. It does not seem to have been derived from any extensive sampling and averaging of body weights or other biometric data. The 150-pound baseline for a standard man changed only very slightly in subsequent decades to 154 pounds, which translated into a more pleasing, round figure in the metric system: 70 kilograms. This standard man appeared in popular treatises on dietetics and in medical and scientific journals until the 1940s.[17]

Resistance to the notion of a single standard cropped up occasionally. An editorial in the *Medical Journal and Record* asked rhetorically, "What two persons in the world are identically alike in physical construction and mental capacity? There are none." Therefore, it concluded it was "well nigh impossible" to establish dietary recommendations based on a standard man.[18] A satirical 1909 piece in an alternative-medicine journal, titled "Man as Found in the Drug Store," proposed creating a "man factory" to manufacture men by combining the precise amounts of various elements required to reach the 154-pound standard: "three pounds and thirteen ounces of calcium," "a chunk of sulphur the size of an egg," 97 pounds of oxygen, and so forth.[19]

Nevertheless, the standard man persisted and became singularly important in the food-distribution recommendations mooted by intergovernmental organizations starting in the First World War. In 1918, the Inter-Allied Scientific Food Commission undertook to determine how many calories each Allied country was producing and how much their populations needed in order to allocate food resources (from the United States to Europe, for example) appropriately. The commission decided that "a man of average weight (seventy kilos, or 154 pounds) doing average work during eight hours a day requires daily food having an energy value of 3,300 calories; and, further, that if the supply of this amount of food became impossible a reduction of 10 per cent. could be supported for some time without injury to health."[20] After the end of the Second World War, the standard man resurfaced in the calorie recommendations that the Food and Agriculture Organization (FAO) of the United Nations published in 1950, though he had a new moniker—"reference man"—and had lost a bit of weight to become 65 kilograms (143 pounds).[21] The 65 -kilogram reference man remained central to the FAO's calorie recommendations into the 1980s.

Francis Benedict did not explicitly adopt the concept of a standard man, but his work helped enshrine Euro-Americans as the norm against which all people's metabolism would be judged. In particular, the equations he devised with the biometrician James Arthur Harris for predicting BMR served this function. Indirect calorimetry, though less resource-intensive and complicated than using a respiration calorimeter, was still time-consuming and required training and specialized equipment to do properly, so Benedict strove to develop an even simpler way to estimate BMR. In 1919, he and Harris derived the Harris-Benedict equations, which could predict an individual's daily heat production, and therefore calorie needs, based on their height, weight, and age alone, without needing to directly *or* indirectly measure the metabolic rate.[22] They worked backward from metabolism data that scientists had collected through indirect calorimetry to create an equation that, when the height, weight, and age data from the original subjects was fed into them, would yield something close to the basal metabolism that had been measured for that person. There was one equation for men and another for women.

These equations were widely used in metabolism studies for decades, and they remain commonly used today. To commemorate the centenary of the Harris-Benedict equations in 2019, the journal *Clinical Nutrition* commissioned an expert group to assess different ways of estimating BMR. The group concluded that the Harris-Benedict equations remain "the best available for the prediction of resting energy requirements in healthy, normal-weight individuals."[23]

From what data did Benedict and Harris derive their equations? By their own account, they examined biometric measurements made by the Carnegie Nutrition Laboratory and other cooperating institutions on a total of 136 men and 104 women.[24] They did not specify the race, nationality, or occupation of these subjects, which suggests that they were mostly or exclusively people whom Benedict and Harris presumed to share their own race, nationality, and class and therefore to be unmarked—not demographically unusual or interesting to either the investigators or their intended audience. Benedict and Harris reported that they strove for homogeneity in the subjects they selected as a sample of "normal human individuals."[25] They considered it a significant enough anomaly that a minority of the male subjects were vegetarians and a minority were athletes that they separated these groups out and ran a separate analysis on them.[26] Had they perceived any of the subjects as differing in race, class, or nationality, we can presume that they would have separated them out in a similar way.

Benedict's purportedly universal tools did not work well for everyone. For example, using the Universal Respiration Apparatus required fixing a clip on the subject's nose that held the receptacle in place for their exhalations. Subjects always required some acclimation to the unfamiliar sensa-

TABLE 1

Deviation in Basal Metabolism of Miao Males from Harris-Benedict Prediction Standard (Kilborn Study)

Subject	Per cent deviation	Subject	Per cent deviation
1	+13.0	13	+24.1
2	+26.3	14	+23.9
3	+21.6	15	+28.3
4	+18.5	16	+ 0.8
5	+15.5	17	+ 8.5
6	+31.8	18	+ 6.2
7	+ 5.2	19	+14.3
8	+ 7.0	20	+ 1.1
9	+28.8	21	+21.2
10	+13.1	22	+ 4.2
11	+17.0	23	+20.4
12	+ 1.7	24	+27.1

Average for all subjects .... +15.8

Figure 1.3 Table from Francis Benedict, "Race: A Factor in Human Metabolism," *Proceedings of the American Philosophical Society* 78, no. 1 (October 22, 1937), 105.

tion of the clip, but Benedict eventually discovered, through research done in China, that for subjects with "flat noses" the clip was so uncomfortable that it was unusable.[27] In addition, researchers quickly began to discover that the Harris-Benedict equations returned predictions more accurate for men than for women.[28] As for subjects of different races, even today nutrition researchers observe that the Harris-Benedict and similar equations have "generally performed poorly" when attempting to predict basal metabolism for non-White people.[29]

This did not prevent the Harris-Benedict equations from being used around the world, including in China. In fact, the Harris-Benedict predictions quickly came to serve as a universal norm from which "deviations" could be calculated, as illustrated in the table in Figure 1.3.

The Harris-Benedict standard, alongside other standards derived from testing the bodies of "Europeans dwelling either in western Europe or in the U.S.A.," thus facilitated a new topic of study: comparative racial metabolism.[30] Research in this area began to flourish in the mid-1920s and continued to occupy scientists through the 1930s. Some researchers insisted that race significantly influenced metabolism and therefore ought to be routinely included as a factor to consider when assessing an individual's predicted metabolic rate, alongside age, sex, and weight. Again, the Carnegie Nutrition Laboratory was in the vanguard. Benedict published articles—sometimes alone, sometimes in collaboration with colleagues—whose titles, such as "The Racial Factor in Metabolism" and "Race: A Factor in Human Metabolism," confidently pronounced this conviction.[31]

Constructing the Chinese Normal

As it happens, Chinese subjects were among the first to be considered as the study of comparative racial metabolism emerged. "Orientals" were an object of particular fascination for scientists engaged in basal metabolism research, and not just for Euro-American scientists but for the Chinese scientists who had trained with them, too. The Chinese Physiological Society, an organization jointly founded by European, American, and Chinese scientists at Peking Union Medical College in 1926, decided within a few months of their founding that the very first research program the society would undertake was to be on basal metabolism.[32] Widespread interest in the "Oriental" metabolism did not, however, produce a robust consensus about whether it differed from the metabolism of others.

The first set of studies, starting in the early 1920s and conducted by scientists of White European descent, suggested that Chinese bodies had an average BMR significantly lower than the bodies of White Euro-Americans.[33] Scrutinizing these studies reveals the messy, paltry data underlying the assertions, however. One of the seminal articles was published in 1925 in the *American Journal of Physiology* and titled "The Basal Metabolism of Some Orientals." This one deserves close examination because virtually every subsequent racial-metabolism study cited it, often as a key piece of evidence. In it, Benedict collaborated with Grace MacLeod from the Teachers College at Columbia University and Elizabeth Crofts at Mt. Holyoke College to report on basal metabolism measurements of Chinese and Japanese students enrolled in American colleges. The authors measured the BMR of nine subjects altogether, seven Chinese and two Japanese women between the ages of 21 and 29 years. They then compared these results to what three standard formulas—each derived from tests done on Euro-American subjects—predicted for women of their height, weight, and age.[34] They found that, regardless of which formula they used, the women tested returned BMR lower than the predicted values. Thus, they concluded, "Oriental women have a metabolism per square meter of body surface 10 per cent lower than the commonly accepted standards for American women."[35]

One problem with this conclusion was that when measured, the BMR of American women *also* consistently fell below "the commonly accepted standards for American women." Scientists well understood how unreliable these standard tables could be when applied to a real group of people, as the historian Andrew Ruis has highlighted. Experts trying to assess the extent of malnutrition among American schoolchildren in the 1920s also found that standard tables did not work well for some southern European immigrants.[36] The authors of the study of "Some Orientals" were clearly aware of these pitfalls in the standard tables and conceded that "all the present standards for women are probably 5 per cent too high."[37] Nevertheless, even after arbitrarily knocking off 5% from the predicted values to correct their presumed skew, the researchers found that the values for the 9 women tested were still several percentage points lower than the revised predictions. It was therefore time, the authors insisted, that "a racial effect in the direction of a lower metabolism be recognized with the Chinese and Japanese."[38]

Resting such a broad conclusion on such a narrow sample might strike some readers as precarious. In fact, the evidence base is even more fragile than the numbers suggest. The "striking variability from day to day" of two subjects' BMR confounded the researchers; for "normal physiological studies," the authors wrote, these two women were not "ideal."[39] They dealt with this variability by simply excluding the higher BMR values measured for these women, on the assumption that higher rates were likely to have been elevated by physical activity or emotional excitement and not, therefore, to be *true* basal metabolism. In addition, in this experiment the researchers measured only "Oriental" women, not any of their White classmates; the concept of a control group had not yet been established as a standard practice. Given the variable performance of respiration calorimeters used in different labs, using the same set of instruments to measure White students' BMR in the same setting might have lent the conclusions more credibility.

Nevertheless, threadbare as it was, this study became a key part of arguments that race influenced metabolism. Researchers from the mid-1920s through the 1930s trotted out this study repeatedly as Exhibit A in the case for low Oriental metabolism. The nine women featured prominently in articles in the *Journal of the American Medical Association*, the *Scientific*

Monthly, the *American Journal of Physical Anthropology*, and the *Proceedings of the American Philosophical Society*, not to mention later issues of the *American Journal of Physiology*. A few studies of similar size and quality—often fewer than 10 subjects, generally without any control group for comparison—joined MacLeod, Crofts, and Benedict's study to buttress the concept of sluggish Chinese metabolism.

Most of these studies acknowledged that too little data had accumulated to draw any confident conclusion about race and metabolism—and then proceeded to draw the conclusion anyway. Herbert G. Earle, an English physiologist working in Hong Kong who was one of the first scholars to raise the idea of low Chinese metabolism, asserted in a 1928 article that the number of female subjects whose BMR had been studied was "too small to warrant discussion." He discussed them all the same, devoting a paragraph to MacLeod et al.'s nine women to support his conclusion that "the Chinese do exhibit a lower metabolism than Westerners." Even 30 years after comparative racial metabolism had first emerged as a topic of study, a 1952 meta-analysis purporting to examine "all available literature dealing with basal metabolic rate" decried the "paucity" of that literature and the internal inconsistency of the data that it contained. Nevertheless, its author concluded that basal metabolism rates differed "for different varieties of man, decreasing in the order American, European, Mongoloid, with the Indian series at the same level as the last."[40]

The conclusion that Chinese and Japanese bodies generated less heat than the "accepted standard" conformed suspiciously well to the researchers' preexisting perceptions about how Japanese and Chinese people differed from Europeans and Americans, as their incidental comments reveal. Earle mused that it might be because they were not immersed in the hectic pace of modern Western life that Chinese people did not burn calories as actively as Westerners. Or perhaps, he thought, Chinese people's more philosophical and contemplative nature translated into lower metabolism.[41] MacLeod, Crofts, and Benedict opened their article by observing, "The general perception has long prevailed that the Chinese and Japanese . . . have lower food needs than Americans and Europeans," because Chinese and Japanese laborers seemed to be able to work hard and productively for a long time on what Western observers considered insufficient meals.[42] H. Necheles, a University of Chicago physiologist sent to China on a Rockefeller fellowship to do metabolism experiments, reflected on how his anecdotal observations of Chinese character might help explain his research findings: "The average Chinese is always more relaxed than the average Westerner; the vivid Southern Chinese as well as the more quiet Northern Chinese. Their muscular tone seemed to be constantly lower than that of Westerners and it was conjectured that this might be the reason of their lower basal metabolic rate."[43]

Such offhand hypotheses required the writers who expressed them to overlook a number of inconvenient facts. The idea that life in 1920s China was slower paced than life in Europe or America, for example, is perplexing. Treaty-port cities such as Shanghai and Tianjin were every bit as frantic and bustling as New York or London, and the overthrow of the last imperial dynasty in 1911 had given way to years of violence and changes in government as different groups vied for power. If chronic stress was thought to factor into basal metabolism, it should have been a significant factor for people in Republican China. Then, too, studies of Indigenous peoples of South America and Southeast Asia had shown significantly *higher* BMR than the "Caucasian standard"—yet their lives were not presumed to be fast-paced and stressful.

Some researchers pushed back against glib claims that "Orientals" had unusually low metabolism. The first to do so were Japanese scientists at the newly formed Imperial Governmental Institute for Nutrition in Tokyo.[44] In an article published in the *Archives of Internal Medicine* in 1926, Seizaburo Okada, Eiichi Sakurai, and Toshio Kameda observed that "there are two different views of the basal metabolism of orientals. On the one hand, it is proved that the basal metabolism of orientals is quite in accordance with that of Western people; others state that a persistently low metabolism exists. It is most important, therefore, to determine this point exactly and whether the Western standards may be applied to the orientals directly or not."[45] Their conclusion, after experiments on 53 Japanese men and women, was that existing standards *could* be applied in Japan: "The basal metabolism of healthy Japanese is quite like that of Americans and Europeans," they wrote, and "no racial difference exists."[46] This reaffirmed a conclusion that their colleague Hideo Takahira had drawn the previous year. Takahira compared the measured BMR of 120 Japanese men and women with the reported BMR of four series of experiments on European and American subjects and concluded that "the basal metabolism of American subjects is essentially the same as that of the Japanese."[47]

The Japanese researchers' work exhibited the same flaws as that of their American and European counterparts: overly broad conclusions drawn from small and incommensurate pools of data. But Takahira's set of experimental subjects was two orders of magnitude larger than MacLeod, Crofts, and Benedict's. And no one could fault the Tokyo scientists for unorthodox methods or substandard equipment; they had trained at the Carnegie Nutrition Laboratory in Boston and used equipment acquired from Benedict himself.[48] Nevertheless, American scientists, when they considered Takahira's work at all, tended either to place it alongside MacLeod, Crofts, and Benedict's nine-woman study as though they were equally meaningful or to dismiss Takahira's subjects as unrepresentative. The American physician Eugene DuBois, for example, in his textbook *Basal Metabolism in Health and Disease*, conceded that Takahira's was "one of the most comprehensive studies of the normal metabolism," yet insisted that "it is hardly fair to average the Oriental figures with those obtained in America."[49]

By the late 1930s, additional research had appeared, and much of it undermined the notion that race was an important factor in determining metabolism. William Adolph, the biochemist quoted at the beginning of this chapter, had been intrigued by the hypothesis in 1930. He then conducted many physiological experiments from his faculty position at Yenching University in Beijing and concluded that climate and diet influenced basal metabolism more than race. He pointed to research showing physiological differences between Japanese and Chinese people resident in East Asia and those who had emigrated to other parts of the world.[50] Similarly, a 1935 article by M. Damodaran and K. G. Ananthanarayanan in *Current Science* signaled its skepticism in its title: "Is There a Racial Factor in Metabolism?" The authors concluded on the basis of their own research that most of the "deviation" from the European norm that Benedict and a colleague had found in a group of South Indian women could be explained by diet, without needing to speculate about a "racial factor."[51] But it was the work of a Chinese researcher in the United States, Chi Che Wang (also known as Wang Chi-lian, 王季蓮, 1894–1979), that most powerfully undermined facile assumptions about Chinese metabolism.

Wang, a Chinese biochemist working at the University of Chicago, collaborated with Jean Hawks, a PhD student in home economics, to measure the basal metabolism of 21 Chinese children born or raised in the United States.[52] All of the children they tested turned out to have *higher* basal metabolism than what the standards tables predicted for their age, weight, and height. The researchers could only conclude that "according to the present investigation the racial influence plays an insignificant role in the metabolism of American-born Chinese children."[53] It was a notably more-nuanced claim than the one that had provoked their study to begin with—that is, the claim that Chinese people had lower metabolic rates than Westerners. Although their research implicitly challenged this assertion, Wang and Hawks did not highlight or attempt to explain the conflict, contenting themselves with remarking how "unfortunate" it was that so few of the studies done on Chinese and Japanese people had included a control group of Westerners.

A couple of years later, Wang was less reticent to take on what she saw as the poorly supported consensus that Chinese people had slow average metabolism. In an article published in the *Journal of the American Medical Association*, she examined all the data that had been published up to that point on American-born girls of Chinese and European descent.[54] By controlling for age and body weight or surface area, major variables affecting metabolism, and applying a variety of standards, she showed that in fact, the data more often showed the girls of European heritage having *lower* metabolic rates than the girls of Chinese heritage. And the data showed larger discrepancies between White girls tested in Boston and White girls tested in Chicago than between White girls and Chinese girls.

Wang's conclusion was, again, framed modestly: "Judging from the foregoing comparisons, it is difficult to conclude that orientals have a lower basal metabolism than occidentals."[55] Still, its devastating implications for the dogma of racial metabolism were clear. She noted, "There is no doubt that there are at present . . . too many normal standards of wide variation, and the danger of comparing basal metabolism data against 'prediction standards' is becoming more and more apparent. In my opinion, it is therefore unsafe to compare the basal metabolism of two groups of subjects, unless the work is conducted side by side or the values are compared with the same standards."[56]

As Wang suggested, most racial-metabolism research suffered from a couple of methodological flaws. One was the incommensurability of standards, the kind of problem that had troubled Benedict and that continued to bedevil nutrition research. There were indeed many "normal standards," or different conversion formulas to estimate basal metabolism based on biometric measurements. The Harris-Benedict equations were the most prominent but not the only ones: There were also the Aub-DuBois, Dreyer, Benedict-Talbot, and Stark methods, to name only a handful. Some estimated BMR from height and weight and some from body surface area; they prescribed different ways of taking the measurements and different equations for converting the biometric data to a BMR estimate. And as Wang pointed out, the different methods could yield quite different results. The Harris-Benedict method tended to produce too-high estimates and not just for Chinese subjects. The method that yielded predictions closest to the actual measured metabolic rates of Chinese and Japanese subjects was Dreyer's, but Dreyer's method never got as much traction as the Harris-Benedict equations. Earle dismissed Dreyer's approach as producing estimates that were consistently about 2.5% lower than measured rates, though he did not make clear why this was worse than a method that created estimates 5–6% too high, as Harris-Benedict's did.[57]

Another flaw was the failure to test subjects of different races at the same time with the same equipment under the same conditions. Given the variability of tools, techniques, and standards in use, as well as the acknowledged impact that climate and ambient temperature had on metabolism, the typical method that metabolism researchers used to compare groups was unsound. They often tested only people whom the researcher perceived to be of one race and then compared those results with published data on people whom the researcher perceived to be of a different race—data collected in a different time and context by a different scientist. Then, too, the very notion of race was fuzzy. Some researchers acknowledged that "Orientals," "Westerners," and even "Chinese" were labels so general as to be almost meaningless; they concealed a huge amount of diversity among the people to whom they were assigned.[58]

Despite these consistent flaws and the well-reasoned opposition that researchers such as Wang mounted to the theory of low Oriental metabolism, Benedict and his collaborators continued to support it in the late 1930s. They were helped, no doubt, by a zeitgeist that emphasized race. Eugenic movements were influential worldwide, including in Asia. Researchers in the Japanese-run puppet state of Manchukuo showed a keen interest in physiological research that differentiated Manchus, Chinese, Koreans, Mongols, and sometimes Japanese and Westerners.[59] This proclivity to see difference meant that even when research that found no disparities featured larger numbers of participants and careful, transparent methods, it was often not treated as decisive or even significant.

In one particularly revealing case, Benedict even discounted his own research results when they seemed to contradict his conviction that Chinese people had a lower metabolic rate than Westerners. By collaborating with Leslie Kilborn, a Canadian physiologist who worked at West China Union University in Sichuan, Benedict was able to access many more Chinese research participants than he had when collaborating with US-based scholars such as MacLeod and Crofts. In 1937 Kilborn and Benedict reported the results of a study they had conducted in Sichuan that tested the basal metabolism of both White and Chinese people at the same time under the same circumstances, thus remedying a major flaw that Wang had observed in previous studies.[60] They found, to their surprise and consternation, *no* significant difference in basal metabolism between the groups tested.

However, rather than entertain the possibility that it might, in fact, be true that there was no inherent metabolic difference in the Chinese body, they scrambled to explain why the Chinese subjects in their experiment could not be considered representative of Chinese people in general. They speculated that the disordered state of politics in Sichuan in the previous few years as compared with calmer places such as Hong Kong and Beiping must have caused Chinese residents there a lot of distress, which raised their metabolic rate.[61] This political upheaval would not have affected White Sichuan residents to the same degree, the scientists argued, given that they were able to leave when they wanted. They also suggested that Chinese residents of Sichuan, unlike European and American residents, mostly lived in unheated houses and had fewer resources such as "woolen clothing and bedding" to protect themselves from the cold, so that might have raised their metabolic rates while those of their White counterparts stayed the same.[62] They recognized, in short, that environmental factors, both climatic and sociopolitical, could affect basal metabolism.

This recognition might have led Kilborn and Benedict to reflect on the mutability of race had they been able to regard their work from a different point of view. Their conviction that Chinese bodies had lower vital capacity than White ones was too entrenched to allow such reflection, however. In the same year, 1937, which also happened to be the year in which he retired, Benedict also published an article called "Race: A Factor in Human Metabolism," which continued to insist that "even when allowances are made for differences in temperament (Oriental placidity versus Occidental tenseness) . . . it is clearly shown that Oriental races in general have a metabolism somewhat lower than that of Caucasians in the United States."[63] To support this conclusion, he cited a number of studies done by European and

American scientists and ignored Takahira's and Wang's work altogether (in fact, the article cited no research by scientists who were not European or American). Summing up the nearly two-decade focus of his work, he wrote, "The Nutrition Laboratory's racial survey has clearly established that there are marked differences in the basal metabolism of human races. The importance of studying the causes of these metabolic differences cannot be overstated."[64]

Every scholar wants to believe that their work is important, but what did Benedict and others engaged in this sort of research think was so vital about understanding basal metabolism and comparative metabolism? Basal metabolism had a medical purpose, for one thing: Knowing how many calories one might expect a healthy body to burn per day could help indicate the presence and severity of a disorder that caused bodies to expend more or fewer calories than normal, such as a thyroid disorder or diabetes. More pernicious were the political and economic uses of comparative metabolism research: Data about the metabolism of different groups of people could help governments—and other organizations responsible for distributing or rationing food—determine how much they thought people really needed.

When Harris and Benedict first published their seminal, standard-setting work in 1919, they described how food shortages among Allied countries in the recently concluded world war had motivated it. "If communities or nations are to be stringently rationed during periods of emergency," they wrote, their leaders needed "to know the minimum amounts of food required to maintain health and efficiency." In peacetime, too, "industrial efficiency and the physical well-being of the population demand exact knowledge of the amount and proportion of the different kinds of food which should be taken by the individual."[65] The scientists at Tokyo's Imperial Government Institute for Nutrition similarly emphasized how such knowledge could contribute to rational governance: "The solution of the minimal caloric requirement of man offers a way to the solution of population, food problems, labor questions, and wage problems."[66] By implication, as comparative racial studies developed, such studies could help inform policy when a government or organization was providing food to people from different racial groups.

Precisely such an occasion presented itself after the Second World War as food aid became an important element of humanitarian aid and Cold War strategy. As the historian George Solt has shown, in postwar Japan, the US occupation government proposed to provide Chinese and Korean residents with 1,800 calories per day, a far smaller ration than the 2,400 calories they provided to White residents. Simultaneously, Allied governments were providing the citizens of occupied Germany with 60% more calories per capita than they were providing to the Japanese. This practice contributed to the tremendous hardship that the Japanese experienced in the first few years after the war. It was implicitly endorsed by earlier nutrition science that had suggested that "Oriental" people had slower metabolisms and therefore could survive on fewer calories than "Occidental" people.[67] It contradicted what Takahira and his colleagues had found in their research, but presumably the work of Benedict and the Carnegie Nutrition Laboratory influenced the Americans in charge of the occupation government more than the work of the scientists at Japan's former Imperial Government Laboratory. Only years after the end of the occupation, in 1960, was Shinjiro Suzuki, then the head of the National Institute of Nutrition in Tokyo, able to quietly change the faulty basal metabolism estimates that had been adopted for "Orientals" in 1949. "Although the standard values for Japanese in the past were considerably lower in comparison with Westerners," he wrote, "the Committee on Calorie Requirements, after investigation of this point, concludes that this difference is due to the erroneous statistical methods applied in the past."[68]

Elsewhere in the postwar years, the concept of racial metabolism was more muted or even explicitly rejected. In 1946, for example, the newly formed Food and Agriculture Organization (FAO) of the United Nations convened a special Committee on Calorie Requirements to develop a common standard for calorie needs that could be applied anywhere in the world.[69] The committee, comprised of eight experts from Europe and the United States and the Chinese biochemist Hsien Wu (Wu Xian 吳憲),[70] decided against adopting any existing scale of calorie requirements, recognizing that those were "based on data obtained in the study of individuals and populations of the Western type, and were primarily intended for use among Western populations." Instead, they chose to create an imagined "reference man" and "reference woman" based on "the human group which has been most carefully studied with respect to anthropometric measurements, physical fitness, food consumption, and energy expenditure, i.e. young men and women living in the temperate zone who are in good health and whose nutritional status is satisfactory. The average age of this well-studied group can be taken as being approximately 25 years, and the average weight approximately 65 kilograms for men and 55 kilograms for women."[71] In other words, rather than create a universal table, the committee provided a recommendation for a standard man and woman and invited local authorities to adapt that recommendation to the population in their charge, using data on that population's average age, height, weight, surface area, and level of physical activity to make the adjustment. In effect, however, by basing the reference man and woman on the "most carefully studied" human group, the committee was making white-collar Europeans and Americans the global norm, so they did not escape the problem of unrepresentative data that they'd decried when discussing older standards.

The committee addressed and dismissed the question of racial metabolism early in their report. "The population of the world is made up of many different races," they wrote. "It can, however, justifiably be assumed that, as far as present knowledge goes, individuals of the same size living in the same environment and having the same mode of living will have the same calorie requirements, whatever their race. In other words, race *per se* does not influence calorie requirements."[72] Apparently Benedict and the other scientists who had urged their colleagues to recognize race as an important factor in metabolism had failed. This may reflect the thin data and dubious methods on which their claims were based. Wang had not been alone in recognizing the flaws. It also likely reflected a change in the intellectual environment. After the horrors of the Holocaust, fascination with racial difference seemed more sinister than it had in the past. The Carnegie Institution of Washington quietly closed its eugenics research laboratory in 1944. As for the Nutrition Laboratory, after Benedict retired in 1937, his program of racial-metabolism research fell by the wayside, and "the laboratory lost its best advocate," in the words of the commemorative history published to mark the Carnegie Institution's hundredth anniversary. The laboratory did some contract work for the US government during World War II and then closed in 1946.[73]

This was not the end of the story for comparative racial metabolism, however. Starting in the 1980s, the FAO committees estimating international calorie requirements changed their methodology in a way that brought racial comparison back into consideration. The first three iterations of the FAO's calorie-requirement reports, published in 1950, 1957, and 1973, derived recommendations from calculations of energy *input*—that is, how many calories a healthy person typically consumed each day. The experts conceded that basing calorie requirements on energy *expenditure* would be more precise but wrote that the lack of data on energy expenditure made that impossible. Not enough was known about BMR and about how different kinds of activities and circumstances affected energy use.

Nevertheless, in the 1985 report, the committee decided that the time was right to make the switch from basing recommendations on calorie input to basing them on calorie expenditure. Accordingly, they dropped the concept of the "reference individual." The growing obesity problem in the populations on which the "reference individual" had been modeled—adult Europeans and Americans—convinced the experts that using their diets as the normal calorie-intake standard was no longer viable. Instead, the advisers looked for BMR data from which to derive calculations of average energy *expenditure* that could be applied worldwide. The scanty-data problem that their predecessors had faced had not subsided; there were still very few reliable studies of basal metabolic rates among people outside of the rich world.[74] Nevertheless, this committee decided that the importance of switching to a calorie-expenditure model outweighed the drawbacks of an unrepresentative sample. They developed a new set of BMR-estimating equations from an old set of studies. In the data set 87% of the individuals represented were European or American, and virtually half of them were Italian. About half of the Italian data was drawn from BMR studies of young military recruits in the 1930s and 1940s, conducted by Italian scientists at such institutions as the Royal Fascist Academy.[75]

One reason the committee was content to allow such a patently unrepresentative group to stand in for all human beings was that, like every previous FAO calorie-requirement committee, this one was convinced that there were no "ethnic differences" in metabolism. They wrote, "A number of studies have attempted to assess the possibility of ethnic differences in BMR but these have failed to identify any differences that could not be related to the nutritional state or possibly to climatic conditions."[76] Therefore, any group of healthy humans could represent the whole, regardless of their race.

Before long, however, the homogeneity of the subjects in the 1985 report's dataset sparked critique. Discomfort with the persistent gap between equation-produced BMR estimates and actual BMR measurements, especially in non-White populations, sparked what biochemist C. J. K. Henry called "a resurgence of interest in BMR."[77] Scientists began to conduct more basal-metabolism studies on people around the world with the goal of creating new predictive equations that would work better for populations unrepresented in the data sets that had shaped the Harris-Benedict and other early equations. Articles no longer referred to "race" and "racial metabolism"; such terms as "tropical peoples" and "ethnic difference" took their place. But the kind of research conducted under this rubric nevertheless resembled the racial-metabolism research of the 1920s and 1930s. Rather than add data from more diverse subjects to the set from which standard human calorie requirements would be calculated, these new studies proposed equations specifically suited to Chinese or Malaysian or other groups, to be applied only to a person or people identified as part of the relevant group, not incorporated into a universal standard. The status quo did not change, in other words. Scientists still wrote as though the Harris-Benedict equation—derived from the bodies and eating patterns of Euro-Americans in the decades before the Second World War—were an acceptably universal standard and searched for a Chinese normal to be applied only to Chinese people.[78]

How will nutrition scientists in the future approach the relationship between race and metabolism? The signs are not promising. In 2023, the US National Institutes of Health (NIH) announced a 5-year, $170-million-dollar study meant to inform "precision nutrition," called "Nutrition for Precision Health." Precision nutrition, touted as the future of nutrition science, aims to develop individual interventions that can "prevent or treat chronic diseases based on a person's unique characteristics like DNA, race, gender, health history, and lifestyle habits."[79] The study promises to use artificial intelligence (AI) to develop and apply algorithms to data from the participants enrolled in the NIH's "All of Us" precision medicine program. The program aspires to create a database of genetic and health data from a million volunteers, particularly those from groups historically underrepresented in medical research; the ideas is to draw on that data to build a more precise understanding of how individual characteristics influence health and disease. Who are these participants so far? As of February 2026, the NIH reported that more than 600,000 people had completed the initial steps to enroll in the program. Of those, a little over half self-identified as White, 15.2% self-identified as "Black, African American, or African," 16.1% self-identified as "Hispanic Latino or Spanish," 3.9% as Asian, and the rest as "more than one race/ethnicity," "other," or "prefer not to say."

This effort seems destined to reproduce some of the flaws of earlier racial metabolism research. Some minoritized populations are represented in larger proportions than in previous research, one of the stated goals of the "All of Us" program, but more than half of the participants are nevertheless White. Will studies based on this sample thus reinforce the impression that White bodies are the norm and that non-White bodies are deviations from it? The racial categories, moreover, are hardly more precise than the categories that scientists used in the 1920s—one could easily substitute "Occidental" for the NIH's "White" and "Oriental" for the NIH's "Asian." Finally, given AI algorithms' proven track record of absorbing, reproducing, and amplifying racist assumptions, the fact that computers will be processing the large volume of data that the project generates does not inspire great confidence.

The early twentieth-century rise of the food calorie and standard means of measuring metabolism made it possible to treat different foods as fungible and to compare with numerical precision how different bodies interacted with calories. This in turn allowed physiologists' "standard man" to emerge and take his prominent place in nutrition-science literature. Not only was the metabolism of European and American men different from and superior to the metabolism of others, scientists such as Benedict implied, but also such men were the standard by which all humanity might be judged, and to which all humanity might aspire.

Internationally prominent scientists in the 1920s and 1930s suggested, based on thin and contradictory data and preexisting perceptions of Chinese character, that Chinese bodies metabolized food more slowly than the standard man. And while the standard man had largely disappeared from view by the 1940s, he continued to haunt nutrition science, becoming the FAO's "reference man" for several decades. Even after the reference man ceased to appear in official publications, the standards built from the bodies of Euro-Americans tested 100 years ago—equations for converting height and weight to basal metabolism, estimates of calorie requirements—have continued to underpin the presumed global norm.

Considering how the standard man's specter has persisted, contemporary studies such as the NIH's "Nutrition for Precision Health" might seem like a welcome move away from the imperialist attitude that there is one standard body and metabolism by which everyone ought to be measured. The study touts its "large, diverse participant group," for example, and declares that "there's no such thing as a perfect, one-size-fits-all diet."[80] But subjects identified as White, the label that has replaced the "Occidental" of Benedict's day, still populate more than half of the database on which the "Nutrition for Precision Health" research is based, so inevitably the studies it spawns will have a much deeper and richer pool of data about one subgroup of humans than about all others. And the subgroup about which NIH scientists learn the most will be the same one that has dominated nutrition science research from the beginning.

Besides, it is sobering to remember that the racial-metabolism research pioneered by Benedict and his colleagues *also* could have challenged the established standard and fostered a more inclusive one. Ultimately, it did not. The data gathered on Chinese subjects were never presented as relevant to creating a new universal standard. Instead, they were presented as a way to arrive at a *Chinese* normal. The same was true of research on the metabolism of other non-Euro-American peoples around the world. Racial metabolism research did not unsettle or challenge the complacent view that the data and equations derived from the bodies of semi-sedentary European and American men were good enough to stand in as a standard for all humankind. And racial-metabolism research was used to reinforce existing stereotypes and prejudices and to justify different treatment of different populations as identified by race. There is little indication that the gene-focused, AI-inflected contemporary version of this research will do better.

Even if genes do contribute something to basal metabolic rates, and even if scientists could detect a consistent small difference in metabolic rates that roughly mapped onto perceived racial boundaries, the costs of researching these phenomena outweighed the epistemic benefit. The anticipated benefit—that scientists would learn something about human physiological variation—never materialized because the research was too spotty and poorly conceived. Even if the experiments had been methodologically airtight, however, it is hard to believe that the knowledge gained would have been worth the costs, which included reinforcing preexisting racial prejudices and providing validation for policies that deprived less powerful social groups of food.

In subsequent chapters we will see that the belief in slow Chinese metabolism was not the only consequence of this Euro- and Americentric way of thinking about diet. Slow metabolism was just one of a number of deficiencies that scientists perceived in Chinese bodies and diets, which they presented as falling short of a universal standard. As with the standard man and his required calorie count, the benchmarks in these other cases also turn out to have been based on European and American models. And as with the Chinese normal, these other perceived deficiencies reinforced stereotypes about the philosophical, passive nature of the Chinese. In the next chapter we will examine the belief that because they depended on rice, Chinese people grew deficient in both vitamins and political power.

2

The Deficient Staple

"Rice Eaters" and the Discovery of Vitamins

In 1901 Myer E. Jaffa, an assistant professor of agriculture at the University of California, published a government report in which he tried to convince his readers that Chinese people ate more things than rice. The US Department of Agriculture had asked him to investigate the nutritional value of, in essence, weird diets. Since the department already had ample data on people who ate a more usual mix of meat and vegetables, Jaffa was supposed to collect information on the outliers, including "a fruitarian colony who claimed to live almost exclusively on a diet of raw fruits and nuts" and Chinese people, "said to live very largely upon rice."[1]

To his surprise, Jaffa found the diets of his Chinese subjects—a dentist's household, a group of laundrymen, and a dozen truck farmers, all California residents—less odd than he had expected. "It appears," he reported, "that many foods besides rice are eaten."[2] Even more surprising was that his subjects' diets appeared to be nutritionally adequate by the accepted scientific standard. The mix of protein, carbohydrates, and fats they consumed and the approximate number of calories met the benchmarks that earlier researchers had established. When it came to diet, he concluded, "The Chinamen seem to have selected a well-balanced one, suited to their wants."[3] This conclusion was so different from what educated American readers expected that Jaffa emphasized it several times.

Why were White Americans in the early twentieth century so surprised that their Chinese neighbors had varied and nourishing diets? One reason may have been the way American propaganda had characterized Chinese immigrants. Labor activists advocating for the 1882 Chinese Exclusion Act, a near-total ban on immigration from China to the United States, claimed that Chinese people could work more cheaply than those of European descent because they could live on poor, monotonous diets that would not sustain White workers. The American Federation of Labor's 1902 pamphlet urging Congress to renew and expand the Exclusion Act used as its subtitle "Meat vs. Rice," alluding to what Senator James G. Blaine had said when arguing in favor of the original act more than two decades earlier: "You can not work a man who must have beef and bread, and would prefer beef, alongside of a man who can live on rice. In all such conflicts, and in all such struggles, the result is not to bring up the man who lives on rice to the beef-and-bread standard, but it is to bring down the beef-and-bread man to the rice standard."[4] The implication was that having to compete with rice eaters would eventually compel American workers to accept a diet as meager as that of their competitors. Not only would this degrade them as individuals and reduce their quality of life, but it would also imperil the nation, since rice-eating nations were supposedly stagnant, despotic, and in decline.

Surprisingly, by the early twentieth century a dark cloud of suspicion had come to hover over rice even among Chinese observers. One of the main reasons was that practitioners of the nascent science of nutrition had found rice guilty of causing a disease that was a visible concern across East Asia at the time: the vitamin B1 deficiency disease known as beriberi. Even though it was not rice itself but modern processing techniques and modern food-provision patterns that caused beriberi outbreaks, the conclusion that eating rice made a person sick fit widespread ideas about the Chinese nation as sickly and weak. The simpler conclusion—that rice is an inadequate staple—gained traction as well.

This chapter explores how a grain important to many Chinese diets came to seem like a hindrance to health and progress. It argues that in the first half of the twentieth century, the new concepts of the vitamin and the vitamin-deficiency disorder absorbed anti-Chinese prejudices and expressed them in more neutral terms. Later, in the decades after the Second World War, the governments of Taiwan and China unintentionally reinforced the idea that rice was nutritionally problematic. Thus scientists and national leaders inadvertently enabled an idea rooted in anti-Chinese prejudice to gain a foothold even in Chinese societies. The story of beriberi and anti-rice arguments is a classic example of an imperialist mindset: perceiving a problem of recent vintage as though it were timeless and characteristic of Indigenous culture and thereby making the habits and technologies of the metropole the obvious solution. In this case, the problem was beriberi, and the perceived solution was switching from rice to wheat.

The Birth of the Food Staple

Jaffa's discovery would not have surprised Chinese scholars. They knew better than to think that Chinese diets could be reduced to one representative grain. A great many Chinese eaters, such as those from northern areas where wheat, millet, barley, and sorghum grew, rarely or never ate rice, and those who did also ate mixed grains and varied vegetables. In any case, the most revered grain in ancient China had been millet; there was no rice equivalent to Hou Ji 后稷 (Lord Millet), the deity in Chinese lore who first taught people how to grow grain.[5] It was not until roughly the fourth through eleventh centuries AD, when migration shifted the center of the Chinese population south to the Yangzi River basin, that rice rose to the fore. This migration took Chinese farmers from the dry loess plateau southeast of the Gobi Desert to much wetter terrain, where they encountered Native peoples growing rice and began to experiment with growing it themselves. From that point forward, rice was prominent in Chinese culture.[6]

Even so, other grains never lost their places in cuisine or dietetic theory. Classical texts insisted that right eating was diverse and balanced. Rice was but one of the foods in the *gu* (穀) category—substantial, plant-based foods that were important to a meal and were neither meat nor vegetable. Different texts included different foodstuffs in their sets of five, six, or nine *gu*, ranging from millet, barley, and hemp to sweet potato and soybeans. In every case, however, the underlying principle was the same: Eating exclusively or even mostly a single type of food misaligned a person with their circumstances and environment and was thus a recipe for ill health. Rice might be appropriate for one season, region, or life stage but not for another; it might combine better with some foods than with others. One eighteenth-century text, for example, warned sick children and pregnant women against eating sticky rice under any circumstances (it also said that the risk extended to cats and dogs) and advised that eating rice together with horse meat induced intractable disease.[7]

Even rice alone, *mi* (米), was a category rather than a unitary object. Dietetic texts parsed the characteristics and health effects of different varieties, describing some strains as warming in nature and others as cooling. Some strains of rice could "harmonize the five *zang* 臓 organs, clear the blood vessels, develop the muscles and flesh, [and] strengthen the tendons and bones," while others could "adjust the stomach, stop diarrhea, cause the *qi* 氣 to descend, resolve agitated thirst, and whet the appetite."[8] Much depended on where the rice was grown, its color and stickiness, and when it was planted and harvested.

In short, rice's place in premodern Chinese diets and dietetics was complex. It was important, certainly, being praised by gourmands, prominent among the *gu*, and grown and eaten in large amounts. Still, in many Chinese diets it was never the most important grain. Nor was it perceived as homogeneous. Chinese medicine specialists understood rice specifically and contextually. By the early twentieth century, however, a new way of looking at rice had arrived in China. Rice's new status as a "staple food" highlighted its sameness rather than its variability and made the grain an abstract symbol of race and civilization.

Today we speak of staple foods as though the term's meaning were self-evident. It is surprisingly imprecise and unstandardized, however. A 2013 volume titled *The Atlas of Food* asserts that "a staple food is one that is eaten regularly and which provides a large proportion of a population's energy and nutrients."[9] The authors do not specify how regularly it must be eaten or how large a proportion of a population's energy and nutrients it must provide to earn the label. The *Oxford Companion to Food*, from 2014, concedes that staple foods and staples "are terms which carry various possible meanings, depending on the cultural and geographic contexts."[10]

Yet this term, so ubiquitous and apparently commonsensical, seems to have developed quite recently—perhaps no earlier than the nineteenth century. The term's distant precursor originated in a fourteenth-century British system that prevented merchants from selling a particular commodity (often wool) anywhere apart from designated towns, known as "staples," both on the British Isles and in continental Europe. By the sixteenth century the word's usage had broadened to mean not only the places where products were sold but also the stuff that was sold there, and by and by it also came to refer to "a principal industrial product of a county, town or district" or "the principal or basic food on which a community lives." The latter two definitions, which are distinct from one another, nevertheless occupy the same entry of the *Oxford English Dictionary*. Although the *OED* entry offers 15 historical examples that begin as early as 1616, almost all refer to a place's principal trade product or commodity.[11] The earliest example it offers of the word "staple" indicating the main thing a group of people eats is from 1971.

Pre-1971 examples of the food-staple concept do exist, however. Consider this from an 1816 article about Ireland published in the *Western Herald* of Steubenville, Ohio: "Potatoes, the *staple food of the poor in Ireland* (we understand from late accounts from that country), have been sold as low as 6 pence per bushel." A May 29, 1822, piece in the *Hampden Federalist and Public Journal* of Springfield, Massachusetts, described corn (maize) as Mexico's primary staple food and manioc as its second.[12] As these examples hint, the term's use often reflected an assumption that the people under discussion had a simpler and less diverse diet than the writer.

This tendency to reduce a community's diet to one primary component developed at the same time that globalization gave some eaters access to a greater diversity of foods than they had been able to consume before. This may seem ironic, but in fact it makes sense given the economic stratification of the time. While more varied produce was appearing in the shopping baskets and on the tables of the well-off, expanding cash crops such as tobacco, sugar, and cotton reduced the diversity of foods that subsistence farmers were producing, leaving many of them with diets that were, in fact, less varied than before.

Summarizing something as complex as a human diet with a single plant mirrored the tendency to ascribe fixed characteristics to diverse groups of people, a tendency that the age of imperialism sharpened dramatically. In the course of this reductionist rush, the staple food that came to be ascribed to Chinese eaters was rice. Wheat, on the other hand, came to be seen as *not* Chinese. Since Anglophone writers played a key role in essentializing both the diets and the people, let us examine how they did it.

Rice Meets Race

As the concept of the food staple was taking shape, the term "rice eater" began to appear as a metonym for peoples of South and East Asia. One of the earliest examples of the English-language term is from 1814, in a book called *The School for Good Living*, which celebrated "the progress of alimentary science, tracing it from the age of pounded acorns to the refinements of modern luxury."[13] The book and its reference to rice eaters could hardly exemplify Western imperialism more clearly. Its author, William Thomas Beckford (1760–1844), was an Englishman born into easy circumstances and thus able to dabble in a number of fields. Besides his book on gastronomy, he wrote a novel, travel narratives, and satires; collected art; sponsored the construction of grand edifices; and served as a member of Parliament. The inherited wealth that made this range of pursuits possible derived from his forebears' positions in British-ruled Jamaica. His great-grandfather had been governor of the colony and his father had made the family's fortune from sugar plantations, worked by an unknown number of enslaved people.[14]

Beckford's family history of domination over people of other races may not have created the dismissive attitude he expressed toward South Asians in *The School for Good Living*, but it likely contributed. Toward the end of his book, having devoted over 200 pages to praising British culinary knowledge and skill, he asserted that missionaries could use such knowledge to "rescue the Hindoos from the horrid obscenities" of their religious practices. He wrote, "What arguments could [the missionaries] use, what means of conversion can they more effectually employ, than by setting before these poor unsophisticated rice eaters the superior excellencies of the European kitchen?"[15] In other words, the easily grasped superiority of British food would help potential converts perceive the subtler, less tastable superiority of the Christian faith.

Beckford may not have coined the expression "rice eaters." It likely was an expression that people in his social circle were already using. Regardless, its use in print expanded over the course of the subsequent century. Its scope and meaning changed, too. In Beckford's formulation, "rice eaters" denoted people of South Asia and connoted a lack of sophistication. British writers in the decades that followed continued to apply the term primarily to the inhabitants of Britain's South Asian colonies, while American writers, less interested in British India, typically used the term to refer to Chinese people. Whereas British writers tended to use "rice eater" as a term of contempt for the presumed passivity of Her Majesty's Indian subjects, American writers more often used it to express fear of Chinese immigrants whom they saw as living more frugally and working harder than American competitors.[16]

Some writers presented rice eaters as weak and pacifistic, citing as evidence the subjection of Indian or Chinese people to Western rule and commerce. As Company Raj gave way to Crown Raj in mid-nineteenth-century India, one writer marveled that "a hundred English wheat-eaters govern as many thousands of the rice-consuming swarms of Lower Asia."[17] As for the Chinese, the American neurologist George Beard was not unusual in calling them "as peaceful and inoffensive as we would suppose a nation of rice-eaters might necessarily be."[18]

Beard's comment provides a telling example of how in racializing diets, writers ignored or erased distinctions, such as those between wheat- and barley-eating northern Chinese and rice-eating southern Chinese. This is not surprising considering that Beard, like most Anglophone writers, had never been to China. Moreover, by the early twentieth century the overwhelming majority of Chinese immigrants to the United States and Europe were from the rice-producing regions of China's far south, which meant that many Westerners remained unaware of the large numbers of Chinese wheat eaters.[19] Westerners with experience in China—missionaries, diplomats, traders—understood that diets differed across its vast geographical extent and large population. But in the popular imagination, most Chinese people ate rice and not much else.

As the examples above show, by the turn of the twentieth century, "rice eater" had become as much a racial or civilizational identity as a description of people who ate a particular commodity. English-language writers had begun to use it as a synonym for "Oriental" or "Eastern" civilization. Accordingly, Western civilization began to require its own dietary metonym, so that writers could contrast the rice-eating Other with the x-eating Self.[20] The one that emerged was "wheat eater." An early example appeared in an 1870 piece titled "Wheat as a Civilizer." The American author declared that wheat eating was essential to success in "the grand march of the race"—meaning the White race.[21] By the final years of the nineteenth century, some Anglophone writers saw a world inhabited by only two kinds of people of any import, defined by diet and race: Western wheat eaters and Asian rice eaters.

In his 1898 presidential address to the British Association for the Advancement of Science, for example, the physicist and chemist Sir William Crookes invoked race as he decried a global wheat shortage. Wheat, he observed, "is the most sustaining food grain of the great Caucasian race, which includes the peoples of Europe, United States, British America, the white inhabitants of South Africa, Australasia, parts of South America, and the white population of the European colonies."[22] Later, he emphasized how devastating an ongoing shortage of wheat would be, not to the many who would likely starve, but more importantly, to the progress of the race:

> If bread fails—not only us, but all the bread-eaters of the world—what are we to do? We are born wheat-eaters. Other races, vastly superior to us in numbers, but differing widely in material and intellectual progress, are eaters of Indian corn, rice, millet and other grains; but none of these grains have the food value, the concentrated health-sustaining power of wheat, and it is on this account that the accumulated experience of civilized mankind has set wheat apart as the fit and proper food for the development of muscle and brains.[23]

"Rice eater" had thus gone, over the course of the nineteenth and early twentieth centuries, from a disparaging description of South Asians with their simple diets to an all-encompassing label imputing racial inferiority—feebleness, backwardness, pacifism, and stupidity. The context of Sir Crookes' words—an address to his country's most eminent scientists, later published in the illustrious journal *Science*—indicates that he and his colleagues were not immune to this kind of popular prejudice. They were surrounded by the same notions about civilizational hierarchy that inspired writers for popular magazines, and those notions influenced the questions they thought to ask about rice, the data they considered relevant, and the way they interpreted that data.

Reinforcing Racism with Vitamins

In the late 1910s, with presumptions about "rice eaters" and "wheat eaters" already firmly in place, a new way of thinking about nutrition began to emerge: the New Nutrition associated with the invention of the vitamin. This major paradigm shift relieved suffering among people with vitamin-deficiency disorders, but at the same time it ended up giving prejudice against rice eaters a scientific veneer. This section examines the history of vitamins from the perspective of nutritional imperialism, arguing that scientists' reductionistic approach to foods reinforced simplistic views about race and diet.

In the first decades of the 1900s, scientific ways of assessing the adequacy of foods contributed to a focus on rice by itself, rather than in combination with other foods as it was almost always eaten in reality. Nutrition scientists, both those in Europe and North America and those in China who had been trained abroad, fed experimental rats constrained diets and judged how well they fared. They used calorimeters and chemical reactions to judge the caloric and macronutrient content of foodstuffs—how much water, protein, carbohydrates, fats, and minerals they contained. This information was only useful, however, if one also knew the ratios of macronutrients that human bodies required to thrive. With those two pieces of information, a person could arrange the foods to match the needs.

The American chemist Wilbur Atwater, whom we met in the previous chapter, began in the 1890s to determine what input human bodies required by placing individuals inside a respiration calorimeter. Other kinds of observations were less controlled. The researcher might observe everything a subject ate over a certain period, or have a subject record or recall what they ate, and then calculate the total caloric and macronutrient values of the foods consumed. These figures then informed dietary guidelines.

In the vast majority of these experiments, the person in the sealed box or writing down his daily diet was a White adult male. Atwater's *Principles of Nutrition and Nutritive Value of Food*, published posthumously in 1916, shows this clearly. A table purporting to show "Food consumption of persons in different circumstances, and proposed dietary standards" implied, by identifying the nationality of the experimental subjects—e.g. "Men (American)" and "Men (German)"—that Atwater considered nationality relevant to dietary requirements (Table 2.1). The numbers in the table, however, made it clear that only a tiny number of East Asian subjects had been included. The data on northern Europeans and on White Americans of northern European descent came from at least 72 different studies, each study representing an unknown number of subjects, but clearly more than one in each case. The data on East Asians came from at most four studies, each study representing only one person, as the descriptions make clear: "Japanese professor," "Chinese dentist in California," "Chinese laundryman in California," and "Chinese farm laborer in California"—all singular. Even so, the dietary standards that Atwater offered based on these studies at the end of his table were nationality-neutral. They differed only in terms of the amount and type of work that a person was doing and whether they were male or female.[24]

According to this early research, rice was one type of food among many, richer in some elements and poorer in others, but not marked as particularly problematic. A 1916 textbook on food manufacture, designed for use in American high schools and colleges, informed its reader that rice was "rich in starch," and "poor in protein, fat and mineral matter." Its poverty in the latter nutrients was not a problem for those who ate rice, however, because they wisely supplemented it with other foods: "In the East the deficiency of protein is supplied by the addition of leguminous plants, a combination of

Table 2.1. "Miscellaneous" Section of Atwater's Table Titled "Food Consumption of Persons in Different Circumstances"

Group	Number of Studies Included in Averages	Actually Eaten Protein	Actually Eaten Fat	Actually Eaten Carbs	Digestible Protein	Digestible Fat	Digestible Carbs	Fuel Value	Nutritive Ratio
Negro families in Alabama and Virginia	39	86	145	440	79	138	427	3395	9.3
Italian families in Chicago	4	103	111	391	95	105	379	2965	6.5
French Canadians in Chicago	5	118	158	345	109	150	335	3260	6.2
Bohemian families in Chicago	8	115	101	360	106	96	349	2800	5.3
Inhabitants Java village, Columbian Exposition, 1893	1	66	19	254	61	18	246	1450	4.7
Russian Jews in Chicago	10	137	103	418	126	98	405	3135	5
Mexican families in New Mexico	4	94	71	613	86	67	595	3460	8.7
Chinese dentist in California	1	115	113	289	106	107	280	2620	4.9
Chinese laundryman in California	1	135	76	566	124	72	549	3480	5.7
Chinese farm laborer in California	1	144	95	640	132	90	621	3980	6.2
United States Army ration, peace	-	120	161	454	110	153	440	3730	7.1
German army ration, peace	-	114	39	480	105	37	466	2725	5.2

Source: W. O. Atwater, *Principles of Nutrition and Nutritive Value of Food* (Washington, DC: Government Printing Office, 1916), 35.

Note: Note the small number of Chinese people included here. Atwater segregated the people in this section from the categories that came earlier in his table, such as "Persons with Active Work," "Persons with Ordinary Work," and "Professional Men." Ethnically unmarked groups such as "Rowing Clubs in New England" dominate those earlier categories.

rice and legumes being a cheaper complete food than wheat and meat."[25] Thus G. S. Fraps, the state chemist of Texas, was able to conclude in the same year that "rice has a high food value. . . . There is no reason why the lower grades of rice should not be used for human food in this country, as they are used elsewhere."[26]

Soon, though, nutrition scientists would change their tune. The blow came with the development of the New Nutrition. Nutrition science up to the second decade of the twentieth century had focused mostly on calories and a few categories of macronutrients—water, carbohydrates, protein, fats, and minerals—and attempted to establish baseline amounts necessary to sustain life and function. As we have just seen, from that perspective rice did not stand out as comprehensively superior or inferior to other grains. By the 1920s, however—which is coincidentally when nutrition science was beginning to establish an institutional presence in China—attention had turned to focus almost obsessively on the newly defined micronutrients called "vitamins" and to engineering diets for flourishing and self-improvement rather than simply subsistence. A key book representing this change was Elmer Verner McCollum's *The Newer Knowledge of Nutrition*, published in 1919.[27]

The puzzle that had led scientists to invent the vitamin was a disease called beriberi, which had become prevalent and apparently intractable in certain settings.[28] Prisoners, soldiers, agricultural laborers, and factory workers, mostly in East and Southeast Asia, were coming down with crippling pain and swelling in their lower legs, which sometimes progressed to mental confusion, heart palpitations, and death. Scientists urgently sought out the cause. They focused first on germs but eventually realized that beriberi had something to do with what the sufferers ate. By the 1910s, they had reached a consensus that it was necessary for humans to regularly consume some trace element unrecognized up to that time—something additional to the categories previously established. Beriberi victims' diets had lacked that "accessory factor," and its absence had induced their symptoms. Researchers quickly realized that the beriberi-preventing factor was only one of potentially a great many such substances. People did not need large amounts of these things, whatever they were, but without them, terrible conditions could develop—not only beriberi, but scurvy, pellagra, and rickets, too. The "accessory factors" that prevented these diseases came by 1912 to be known collectively as "vitamines," a word that before long was amended to "vitamins" to clarify that not all things in this group shared a common molecular structure as did all amines.[29]

Rice played an important role in establishing this new category so central to later nutrition science. After all, when beriberi emerged, it frequently did so from diets comprising mostly rice. The beriberi-causing diets had another thing in common that attracted less comment: In many cases, their eaters inhabited an institutional setting in which those doling out the food strove to minimize cost. Whether it was plantation or factory owners devising rations for laborers, army planners determining provisions for soldiers, or the directors of hospitals, prisons, or boarding schools deciding what to put on the menus for people in their charge, all wished to spend only as much as was necessary to supply adequate nourishment. This translated to meals with little variety, meant to satisfy needs for calories, protein, carbohydrates, and fats. Earlier nutrition-science research had convinced authorities in charge of canteens and provisions that they could design nutritionally satisfactory diets by balancing macronutrients, but ironically, the diets they designed induced vitamin-deficiency disorders.

A 1937 report by the Preparatory Committee of the Intergovernmental Conference of Far-Eastern Countries on Rural Hygiene, a group convened by the League of Nations, underlines this point. The committee observed that beriberi was "localised mainly in Asia." They went on, however, to emphasize that even there it had "never been widespread among rural populations; it is connected rather with towns, armies, shipping, prisons, schools and large undertakings—in short, places where catering is collective."[30] The key element, in other words, was not Asian diets but institutional diets.

It did not help that in such diets, increasingly the rice offered had been steam milled, a new industrial mode of processing that removed the bran sticking to the grain much more effectively than previous methods had done. That made the rice lighter and cheaper to transport, and white rice kept longer because it was less inviting to insects and microbes. At the same time, though, steam milling made rice less nutritious. The bran that the mill polished off so completely was the part of rice that contained beriberi-preventing vitamin B1. Economy was the deadly watchword.[31]

It was easy to ignore these contextual details and notice only one commonality: rice. This is not surprising since beriberi had been perceived as a disease of rice eaters even before scientists understood vitamin deficits. As one observer, himself a "wheat eater," put it in 1906, "It is pretty well agreed that this paralytic disease comes from rice eating exclusively. We wheat eaters are wholly free from it. To prevent this painful and often fatal malady wheat is one of the medicines."[32] In China's case, such observations about beriberi only reinforced popular notions that were growing more entrenched: that China was the cradle of disease, the "land of famine," and the "Sick Man of Asia."[33] The taint of civilizational inferiority that rice had already acquired in the societies that most nutrition scientists inhabited began to infect their work.

Thus the scientists of the 1920s and 1930s ceased to view rice as nutritionally rich in some aspects and poor in others. It became, instead, "a relatively poor cereal at best," in the words of Major-General Sir Robert McCarrison, an Irish doctor working for the Indian Medical Service.[34] In a lecture delivered to Britain's Royal Society of Arts, McCarrison used rice to show how "faulty diet" could jeopardize the health of a nation or a race. Similarly, Elmer McCollum, one of the founders of the New Nutrition, proclaimed rice categorically wanting in a series of lectures he gave at universities across the United States. He declared, "Rice is deficient in all known vitamins, protein, and in several inorganic elements. . . . Owing to the multiple deficiencies of the rice diet it is not possible to decide as to the specific effects of deficiency or lack of a single dietary component."[35]

Because nutrition scientists were so focused on the narrow problem of how human bodies developed vitamin deficiency diseases, they mostly missed or ignored the social and economic factors that produced beriberi; instead, they targeted rice. This reinforced an older, racist stereotype about rice eaters being weak and sickly because they'd chosen the wrong food. Thus, when the first generation of Chinese nutrition scientists went abroad to earn their degrees, they found themselves in a setting where both popular and scientific discourse framed rice eating as problematic. A main message undergirding their field of study at this time was that a rice-eating nation that wanted to become more robust and successful needed to both move away from rice as a staple and improve the quality of the rice that its people ate.

To some degree, Chinese scientists accepted and propagated the idea that rice was nutritionally flawed. They participated actively in the research agenda of the New Nutrition and its ever-finer accounting of the micronutrients needed to flourish. One of the first Chinese textbooks on nutrition science, Zheng Zhenwen's *Nutritional Chemistry* (*Yingyang huaxue* 營養化學) published in 1929, was in effect a book about vitamins; six out of its eight chapters centered on them.[36] Vitamins also featured prominently in the articles published in the *Chinese Journal of Nutrition* and the *Chinese Journal of Physiology* in the 1930s and 1940s. In fact, judging from a bibliography compiled by the biochemist Zheng Ji (Libin T. Cheng) in 1944, vitamins were the most common single focus in early Chinese nutrition research, accounting for nearly a quarter of all nutrition-related publications (Table 2.2).

In the vitamin studies, researchers assessed the nutritional adequacy of Chinese diets in a few ways. One of the most common was to compare the vitamin content of foods bought in Chinese markets against international standards. Another was having subjects consume what scientists deemed a "typical" diet and then measuring the vitamin content of their urine. In at least one study, the scientists sampled the livers of patients who had died at Peking Union Medical College to assess vitamin A levels in their bodies.

Table 2.2. Nutrition-Related Publications by Chinese Scientists, 1927–1943

Category	Subcategory	Number of Publications	Category Subtotal
Protein chemistry		46	46
Fat chemistry		8	8
Carbohydrate chemistry		8	8
Food analysis		37	37
Nutritive value of Chinese foods		36	36
Nutrition			89
	Nutrition: General	6	
	Nutrition: Nutritional technique	4	
	Nutrition: Protein in nutrition	4	
	Nutrition: Mineral in nutrition	17	
	Nutrition: Chinese dietaries	29	
	Nutrition: Vegetarian diets	20	
	Nutrition: Infant feeding	9	
Vitamins		81	81
Digestion and metabolism		50	50
TOTAL		355	355

Source: Based on the bibliography in Zheng Ji 郑集, *Zheng Ji ke xue wen xuan* 郑集科学文选, *1928–1992* [Selected Scientific Works by Libin T. Cheng, 1928–1992] (Nanjing: Nanjing University Press, 1993), 57–72.

The results of this research were far from conclusive. Some foods studied appeared to be nutritionally rich and others nutritionally poor, and some articles concluded that the bodies studied contained adequate amounts of the vitamin in question while others said the opposite. Nevertheless, although fragmented data made it difficult to patch together a general picture, a sense of pessimism emerged, as evidenced by the outpouring of 1930s articles for broad audiences that centered on a "problem" related to Chinese people's diets: "The Eating Problem of My Country's People," "Chinese People's Nutrition Problem," and "Citizens' Nutrition Problem," for example.[37] One 1936 article by Liu Ruiheng, titled simply "The Nutrition Problem," offered the following as evidence of the poor state of nutrition in China: the percentage of middle and high school students found to be suffering from malnutrition (14%) in an unnamed study, the poor results of China's competitors in the recent Olympics, and the number of Guomindang soldiers suffering from beriberi in a recent campaign against the Communists. Examples of malnutrition in factories, prisons, schools, and among professional people, the author added, were so numerous that "I cannot list them."[38]

By highlighting how patchy his data were, I do not mean to suggest that Liu was manufacturing a problem that didn't exist. Given the extreme stresses of wars and famines that devastated China in this period, many people there could not eat plentiful, well-balanced diets. But nutrition scientists' methods were too finicky, their tools too narrowly focused, to build a coherent description of a nationwide nutrition crisis, much less propose solutions for it. As Liu complained in his article, although China did have a few nutrition scientists, "What they mostly do is experiments in laboratories."[39] The artist responsible for a 1936 cartoon titled "The Nutrition Problem" offered a similar critique (Figure 2.1). In the cartoon, a corpulent, well-dressed hospital director peers down from his desk at a group of men in ragged clothes and declares that nutritional deficiency disorders have made them "as skinny as monkeys." "You should eat some milk, eggs, and fresh fruit," he advises. "Oh!???" The men reply, as if the idea of eating well had never occurred to them.

Vitamins, as these observers clearly knew, were not the root of China's nutrition predicament. The New Nutrition's ever-finer focus on trace elements and single foods drew attention away from the bigger picture: Many people, particularly in north China, didn't have access to enough food, or enough varied food, to thrive.

Figure 2.1 "The Nutrition Problem," by Huang Renzhi 黃任之. The caption above the cartoon says: The Nutrition Problem. Hospital director: "All of you lot suffer from nutritional deficiency disorders, so you're as skinny as monkeys. You should eat some milk, eggs, and fresh fruit." Patients: "Oh!???" *Shanghai manhua* 上海漫畫 [Shanghai cartoons] no. 2 (1936): 9.

The scientists themselves recognized the limitations of their work and occasionally expressed frustration about it. In his report to the Fourth International Congress on Tropical Medicine and Malaria in 1948, Hsien Wu (Wu Xian 吳憲) ticked off a number of ways in which "the Chinese diet [was] defective": "deficient in high-quality proteins, low in calcium and vitamins A and D, and frequently also in thiamine, riboflavin, and ascorbic acid." He then pointed to a handful of scattered studies reminiscent of Liu Ruiheng's hodgepodge, including studies of refugees in Shanghai, Chinese troops, Chengdu medical students, and residents of a tuberculosis sanitarium, and concluded that the "general picture of nutritional deficiencies in China is clear and convincing." But remarkably, he spent the remainder of the report explaining why knowledge of specific deficits could not actually improve China's overall condition. Analyzing specific foods and diets—the kind of work that had engaged him throughout his career—was useless under prevailing circumstances. "There is no mystery about oriental foods and diets," he wrote, "nor much hope of nutrition research by-passing social-economic laws. Not until more food is made available per capita . . . does there seem to be any solution of the problem of malnutrition in China."[40] In the meantime, the scientists tackled what *was* in their purview: nutritional deficiency disorders and specific foodstuffs.

On the whole, Chinese nutrition scientists' treatment of rice was more measured than what we have seen coming from American and European scientists. They did not condemn it as a "poor cereal"; they frequently observed that processing technique affected nutritional value as much as a grain's innate qualities and that refined wheat flour was no more wholesome than white rice.[41] Nor did they advocate that people who ate rice switch to wheat instead, generally speaking—although in some instances that reflected their awareness of how hard changing eating habits could be rather than their confidence in rice's value. In 1940 a group of Shanghai scientists lamented, for example, that while millet and whole wheat were more vitamin-rich, "south of the Yangtse rice is the accepted cereal food, and no amount of advice is likely to change such a custom."[42]

Despite such relative even-handedness, it is easy to come away from Republican-era scientists' publications with the impression that rice, itself, causes disease. In part this was because in the 1920s, as the New Nutrition was forming, the primary way scientists induced beriberi experimentally was by feeding animals polished rice alone. In *Nutritional Chemistry (Yingyang huaxue* 營養化學), Zheng Zhenwen explained the procedure. He wrote, "If you want to test the power that a particular substance has to cure beriberi, you should first use white rice to feed a turtledove and make it manifest the symptoms." Having successfully induced beriberi, you then gave the bird whatever "particular substance," often wheat meal, you wanted to test as a potential cure.[43] It is not hard to see how this method—feed an animal rice to induce beriberi, then feed it wheat to try to cure it—might have supported the assumption that rice causes disease while wheat cures it. Scientists presumably understood this approach as a stopgap, a way of researching vitamins until more precise methods were developed. Newspaper articles, however, invoked such animal-feeding experiments to support a broader indictment of rice. As an article in the influential Shanghai newspaper *Shun Pao* 申報 (known in English as the *Shanghai News*) informed the reader, "Using rice exclusively to feed various animals leads them to die within a few weeks; if you feed them wheat exclusively, they stay alive."[44]

It was not only popularizers oversimplifying experts' methods and results who contributed to rice wariness, however. Experts, too, sometimes indulged in generalizations that exaggerated the implications of their work and reinforced the anti-rice message. Consider the way Zheng Zhenwen wrote about beriberi: "We Asians who take white rice as a staple (*yi shi bai mi wei zhu* 以食白米為主) often get a particular kind of disease. This is a phenomenon that since ancient times Eastern and Western scholars alike have paid attention to; as for the cause of this disease, for a long time they couldn't figure it out."[45] The authors of a 1934 article in the *Chinese Medical Journal* similarly described beriberi as a timeless feature of southern Chinese life, induced by rice: "Beri-beri has long been known to be endemic in China, particularly common in the southern part of the country, where rice is the main staple of diet. It is a disease . . . related to the nature of the diet consumed by the people."[46]

Such representations reinforced the imperialist narrative that because they relied on rice, Chinese people—southerners, at least—had always suffered from beriberi until Western science came along to enlighten and heal them. Popular magazines amplified such narratives. A 1946 article in the Chinese magazine *Household Medicine* (*Jiating yiyao* 家庭醫藥) celebrated an American and an English chemist who had developed a new processing method that made coarse rice—known to be more nutritious than white rice—tastier and nicer looking. The author marveled, "Consider that in the world there are about two billion people, of whom about half rely on eating white rice to live, but the method for improving coarse rice succeeded because of the experiments of two bread-eating people (*chi mianbao de ren* 吃麵包的人). Isn't that interesting?"[47] The author thus recapitulated the invented dichotomy between rice-eating Chinese and wheat-eating Westerners and depicted the latter as saviors of the former. Chinese people had needed bread eaters, apparently, to show them how to properly prepare their own principal food.

The scientists were not trying, of course, to make Chinese people look inherently inferior or ignorant or to reject Chinese foodways. In fact, often their efforts were predicated on the idea that differences between Chinese people and Westerners were temporary, changeable, and a reflection of poverty rather than of any innate shortcoming. But the way they framed the issues inadvertently reinforced the message of nutritional imperialism. The contrast between the smallness of their research and the expansiveness of their conclusions is striking. Studies of rats, molecules, and anywhere from a handful to several hundred people informed conclusions about the fate of the Chinese nation. They frequently presented their work as meaning more than it reasonably could.

But linking vitamins with national flourishing was not an insincere or primarily self-interested effort to get more funding or respect. The scientists devoutly wished that their work had greater implications. If one considers, for example, why Chinese nutrition scientists rarely discussed rice without discussing beriberi, a cynical interpretation is to say that it was a way of making their discipline seem indispensable. Unlocking the mystery of beriberi and, by extension, all vitamin deficiency disorders was the signature triumph of their discipline, so boasting about it helped underline the field's significance and their own. But a fairer analysis might conclude that the smallness of what the scientists were able to grasp and do dissatisfied them, especially in comparison to the size and complexity of the problems that their society faced. While the challenge of rescuing China was so unwieldy that it could evoke despair—as Hsien Wu's comments to the Fourth International Congress showed—enabling their compatriots to identify and treat cases of beriberi was one specific, perceptible way in which nutrition scientists *could* improve Chinese lives and futures.

Inadvertently, however, by making beriberi and other vitamin deficits seem like longstanding, nation-imperiling problems, the scientists bolstered the view that Chinese foodways, not just transient economic circumstances, produced malnutrition. What scientists such as Hsien Wu intended to communicate was that Chinese people needed access to more and better food. What they sometimes ended up communicating was that Chinese people had chosen the wrong foods.

Postwar Suspicion of Rice

So far, we have seen how the first scientists to study rice's nutritional value did so in an environment fraught with prejudice about "rice-eating nations." We have seen how their fascination with vitamin-deficit disorders reinforced the idea that rice was deficient. We have also seen how Chinese scientists perpetuated the myopic focus on rice despite knowing that tinkering with individual foodstuffs could not meaningfully address malnutrition. Under adverse circumstances they were doing the doable.

After the conclusion of the Second World War, China's internal political dynamics and international relations changed significantly, and one might have expected the pathologization of rice eating to have waned. Decolonization was the order of the day; with the founding of the People's Republic, China meaningfully stood up against foreign imperialism. Beriberi, by now well understood and easily preventable, began to recede as a public health concern. Nevertheless, rice anxiety persisted. Scientists and national and international leaders continued to scrutinize the grain and advance narratives about its inadequacy. They continued to present it as a prime suspect in the nutritional problems of Asia. Here, we will explore that surprising persistence.

The newly hatched intergovernmental organizations of the late 1940s devoted significant attention to rice. The Food and Agriculture Organization (FAO) of the United Nations convened a Rice Study Group in India in May and June of 1947 to tackle what they called the "rice problem . . . in all its aspects."[48] Although the report they published did not articulate what the problem was, reading between the lines suggests that it was widespread hunger. People whose diets had come to include a large proportion of rice had not been getting enough to eat. Indeed, the 1940s had seen many grim episodes of starvation in South and East Asia. In 1943, for example, a famine ravaged Bengal; 1944–1945 were the years of the Great Vietnamese Famine; in Java it has been estimated that nearly 6% of the population may have died of malnutrition-related causes between 1943 and 1945; and in 1946 and 1947 hunger wracked Japan.[49]

This was not, it seems obvious to say, a rice problem. Yes, rice constituted a significant part of the diets of people in each of these areas. Most had, however, also experienced the severe dislocations of war. It is telling that although hunger likewise pervaded European countries in the postwar period, their suffering did not prompt the formation of an analogous Wheat Study Group. Another significant commonality among the starving people of South and East Asia was that they had all been under colonial occupation at the time of their famines: Bengal was a British colony; Vietnam and Java were under first French and Dutch and then Japanese rule; and Japan was occupied by the United States. Food distribution had to some degree been controlled by a distant metropolitan government.

The authors of the report of the Rice Study Group did not mention the colonial context of food shortages. Instead, they focused on improving the grain itself, detailing how rice-producing countries might better produce, store, and transport it. For example, they could develop better irrigation and drainage methods and apply better fertilizers and seeds, improve soil fertility and pest control, use machines more extensively and improve "the nutritive value of rice."[50]

The nature of colonial relations became starkly visible nonetheless in a disagreement among the Rice Study Group's nine delegations. Five—Burma, India, China, the Philippines, and Siam—were from Asian countries, two of which (Burma and India) were not yet formally independent. The other four—the Netherlands, the United Kingdom, the United States, and Australia—were from countries whose majority populations were of European descent. The summary essay at the beginning of the report, written by Chairman S. Y. Krishnaswamy from the Indian delegation, highlighted an observation some members had made about the composition of the group. "The present Rice Study Group," they had observed, "consists of nine nations, of which four come from Europe and America which produce 5 percent of the world total of rice, while the remaining five members come from Asia which produces and consumes more than 90 percent of the rice. An overdue representation has thus been given to nations whose interest is comparatively small." They had also noted how unreasonable it was that "the administration of rice now lies in the hands of a body situated in Washington [DC, where FAO headquarters were at the time] and dominated by countries which are little concerned with rice and which are hardly competent to deal with the subject, as compared with the countries in Southeast Asia whose whole economy revolves around rice." Going forward, therefore, those members recommended that the main intergovernmental advisory body on rice issues be geographically located in Southeast Asia and constituted mainly of "the major rice-producing and rice-consuming countries of the world."[51]

This view drew strong objections from some delegations, and the dispute was acute enough that the chairman felt obliged to append a statement from each delegation expressing their opinion. The Dutch and UK delegations asserted that rice concerned their countries, too. The UK delegation wrote, "Rice is a world commodity and any organization set up to advise on rice, or control the allocation of the world's surplus of rice, should be a world organization and not a regional one." The Dutch delegation concurred: "The rice problem is of a world-wide character, not only because some countries in the Western Hemisphere have an important interest in the production and consumption of rice but also because there is an interrelationship between rice and other cereals."[52] The American and Australian delegations likewise supported the status quo.

In the end, the Europeans and Americans succeeded in maintaining a prominent place for themselves in international conversations about rice. The study group's summary relegated the suggestion of a Southeast Asia–centered advisory body to the status of a dissenting view. The report stayed silent about prevailing power relations and represented hunger in rice-eating regions as a technical problem that could be fixed by science and engineering: smarter farming, better infrastructure, healthier rice. It obscured the role of Europeans and Americans in creating malnutrition in Asia while emphasizing their role in relieving it—a clear example of nutritional imperialism at work.

A report by the first director of the FAO's Nutrition Division, Wallace Aykroyd, illustrates the consequences for the reputation of rice. The son of a British father and an Irish mother, educated in Dublin, Aykroyd had a critical view of the beleaguered grain. His MD thesis and early published work had focused on beriberi, and when he took the reins at the Nutrition Division he had just completed a decade-long stint as director of the Nutrition Research Laboratories in India. There, he succeeded Robert McCarrison—the doctor who had judged rice "a relatively poor cereal." In this, he shared McCarrison's view. The laboratories' major scientific work under his leadership was an inquiry into *The Rice Problem in India*, as the title of their 1940 report described it.[53]

In 1948, in his new capacity as the head of the world's most prominent intergovernmental organization devoted to nutrition, Aykroyd published a report titled "Malnutrition and the Rice Problem." There, he recapitulated the rice-versus-wheat trope that had circulated since the late nineteenth century. He asserted, "In the rice-eating regions of the world the general level of health—the term 'health' being used here in its widest sense—is low compared with levels attained in western countries in which wheat is the principal cereal."[54] Beriberi naturally got a mention. To emphasize what an ongoing problem this nutritional deficiency disorder was in "rice-eating regions," he observed that in the Philippines beriberi was worse in 1946 than it had been in 1938. He did not think it worth remarking that the Philippines had been under US and Japanese rule during those years; the relevant fact was that rice was the staple of diets there. Nor was it just beriberi that rice diets threatened to induce. In addition, he wrote, "numerous other deficiency diseases and states such as keratomalacia, stomatitis, glossitis, cheilosis, oedema, and hepatic cirrhosis occur more frequently among rice-eating than among wheat-eating populations in general."[55] The solution to the malnutrition brought on by rice eating was twofold, according to Aykroyd: Rice itself needed to be improved through enrichment and better processing, and "rice eaters" (his term) needed to eat less rice and more of other kinds of foods.

This set the tone for rice globally during the crucial postwar years when agricultural systems were being rebuilt and new patterns of trade and aid established. As the Dutch delegation to the Rice Study Group had noted, even a nation that was not a major rice producer or consumer might take an interest in rice shortages elsewhere, thanks to the "inter-relationship between rice and other cereals." Such an interrelationship meant that a nation that produced *other* cereals might be able to sell them to one experiencing a rice shortage. This was precisely the case for the United States in the late 1940s, which found itself with a glut of wheat and a saturated domestic market. In a story that the historian George Solt has vividly told, food shortages in US-occupied Japan threatened to fuel the growth of the Communist movement there, and the US government perceived an opportunity to solve two problems at once. By purchasing excess wheat from American farmers and loaning Japan the money to buy it cheaply, the state could simultaneously alleviate the American wheat surplus and Japan's food crisis.[56] American wheat began to pour into Japan.

The proportions of rice and wheat in Japanese diets began to change correspondingly.[57] The prevailing nutritional wisdom about the poor value of rice likely helped facilitate the shift. Leading Japanese scientists and doctors criticized rice in a strident tone, drawing wheat-rice contrasts every bit as cartoonish as the ones we have seen from nineteenth-century imperialists. A professor of medicine at one of Japan's most prestigious universities wrote in 1958, "When one eats rice, one's brain gets worse. When one compares Japanese to Westerners, one finds that the former has an approximately twenty percent weaker mind than the latter. This is evident from the fact that few Japanese have received the Nobel Prize. . . . Japan ought to completely abolish its rice paddies and aim for a full bread diet." A year later, the director of the Nutrition Bureau in Japan's Ministry of Health published an essay in which he claimed, "The character [*seikaku*] of rice-eating peoples and the character of wheat-eating peoples are naturally different. . . . The former are resigned and passive, the latter are progressive and active."[58]

Not long after the United States brokered the food-assistance program with Japan, the Cold War in Asia intensified. The Chinese Communist Party defeated the Guomindang regime in China's civil war in 1949. The Guomindang government moved to the island of Taiwan, and within a few years its American ally had proposed a wheat-buying arrangement with Taiwan similar to the one established in Japan. In 1954, the United States passed Public Law 480, a scheme by which the government could sell abroad surplus American agricultural products. Among the first targets of this effort was Taiwan.[59]

Taiwan's government was eager to purchase cheap wheat from the United States. They hoped to convince their own farmers to exchange their rice for wheat so that the government could sell Taiwanese rice abroad to bring in more foreign currency. The initial agreement to purchase 123,000 tons of American wheat for the equivalent in Taiwanese currency of nearly $10 million was arranged in 1956, just two years after the passage of Public Law 480.[60]

The government's vision quickly ran into an impediment, however: People in Taiwan ate a lot of rice and hardly any wheat. In a survey conducted in 1956, the ratio of staples in Taiwanese diets was 76% rice, 23% sweet potatoes, and only 1% wheat flour. For the scheme to work, its backers would need to foster a larger market for the tons of wheat flour that would soon start arriving on the island's shores. They thus set out to popularize the less-familiar grain, sending "traveling lecture groups" out to teach

Figure 2.2 Li Kwoh-ting 李國鼎, Minister of the Economy (center), tries donuts at the Wheat and Barley Products booth at Taiwan's 1964 Economic Development Expo. Record number A325000000E/0053/0011/1, National Archives Administration National Development Council 國家發展委員會檔案管理局.

housewives how to use wheat flour in their cooking and erecting wheat-promotion stands at national exhibitions (Figure 2.2).

Although it is clear in internal government documents that the main rationale for buying and promoting American wheat was economic, in public-facing settings the government emphasized nutrition. In a 1962 announcement inviting housewives to apply for the lecture classes, the Wheat Foods Popularization Committee (*mianshi tuiguang weiyuanhui* 麵食推廣委員會) wrote that its main goal was to "improve citizens' nutrition and to promote citizens' health" by encouraging wider use of wheat.[61] A 1968 memo urging farmers to participate in the wheat-rice swap made the case much more explicitly. The first three reasons it gave were all nutritional: "If human beings preferentially eat white rice, that can easily lead to beriberi, stomach disease, high blood pressure, heart disease, kidney disease, and other diseases. . . . The nutrition in wheat flour is greater than in rice, . . . so the physiques of people who eat wheat are all comparatively tall and strong." Therefore, it concluded, "from now on we should start step by step to eat less rice and eat more wheat."[62]

From this perspective, trading one's rice for government wheat was not so much a self-interested transaction as a civic duty, a contribution to public health. Although the Taiwanese officials did not recapitulate caricatures of rice eaters the way their Japanese counterparts had done, their assertions that rice caused disease, short stature, and weakness show how those old prejudices could support a new project. The sudden availability of cheap wheat, combined with the government's pro-wheat propaganda, seems to have had an effect. By 1968, wheat flour had displaced sweet potato, and in the present the percentages of rice and wheat flour in Taiwanese diets are about 55% and 45%, respectively.[63] The majority of Taiwan's wheat, even today, is imported from the United States.[64]

A Brief Challenge to Nutritional Imperialism

In the early People's Republic of China, where leaders frequently denounced imperialism and racism and urged scientists to reject theories associated with capitalist thought, one might have expected nutritional imperialism to have come under attack. And indeed, one does not find in Maoist China the kind of simplistic wheat advocacy that we have seen in Japan and Taiwan. In the post-socialist era starting in the late 1970s, however, rice wariness reemerged. Let us briefly examine that story.

In the New China, vitamin science's imperialist associations did not go unnoticed. The author of one lively essay published in *Popular Science* (*Dazhong kexue* 大眾科學) in 1950 reframed the story of how the Dutch scientist Christiaan Eijkman had discovered the relationship between white rice and beriberi. Eijkman was one of nutrition science's heroes. While working in the Dutch colony of Java in the 1890s, he had observed that when fed polished rice, chickens developed symptoms like those of human beriberi sufferers. His observant eye and clever experiments earned him admiration and a Nobel Prize. In the *Popular Science* essay, however, the author Chen Xinqian embedded Eijkman's work in its colonial environment and highlighted the power relations beneath the experiments.

Chen began, "Long ago in 1897, the Dutch imperialists ruled their Eastern colony—today it is national territory resolutely defended by the Indonesian people, Java." Eijkman then entered the story, not as a brilliant researcher but as an unnamed "medical official in the colony" who "worked in a jail . . . that was a tool for cruelly killing Indonesian people." Eijkman observed that chickens fed prisoners' white-rice leftovers came down with beriberi-like symptoms; this "inspired a research interest for him and the people around him. After going through an appropriate period of research (using prisoners as experimental objects), they discovered: white rice lacks a kind of nutritional component, and this lack causes chickens (and prisoners) to get beriberi; if they changed to feed the chickens (or prisoners) coarse rice, the disease was cured."[65] The parenthetical asides that Chen added turned Eijkman the scientific pioneer into a colonial officer experimenting on Javan chickens and humans with the same disinterested gaze.

Chen also rejected the approach that some capitalist nations had adopted to enrich too-processed grain, which was to fortify white flour with vitamins. "We don't need to and shouldn't study the wasteful manner of the Americans and grind up vitamins and add them in," he declared. "We can more straightforwardly preserve vitamins and the rest with standard rice and flour."[66] Promoting standard rice—not switching from rice to wheat, as governments elsewhere urged—was China's 1950 solution to the known problem that *polished* rice was vitamin poor. Also called 9–2 rice (*jiu er mi* 九二米), it combined 100 kilograms of coarse rice with 92 kilograms of white rice. It was a low-tech way to preserve the vitamin and mineral content of the grain while sidestepping the problem of its unrefined version—namely, its unpalatability.[67]

This critical perspective notwithstanding, in the PRC's early years nutrition scientists continued to take an interest in the relationship between rice and beriberi. The main professional journal, *Acta nutrimenta sinica* (*Yingyang xuebao* 營養學報), published studies analyzing the vitamin B1 content of different varieties of rice and compared its digestibility unfavorably with that of wheat.[68] In a state-of-the-field article published in 1957, Yang Enfu, a researcher in the nutrition department of the Chinese Academy of Medicine and Science, worried that as the country grew richer, more and more people would reject low-status but more-nutritious mixed grains in favor of white rice, perversely leading to more-pervasive beriberi nationwide.[69]

Yang needn't have worried—at least, not about beriberi. By the end of the decade the agricultural collectivization forced by the government's Great Leap Forward economic plan had induced a famine that cost tens of millions of lives. The country's main diet-related health problem was that people lacked the necessary calories to maintain life, not that they were choosing vitamin-poor foods over vitamin-rich ones. Even after the famine subsided in the early 1960s, neither rice nor wheat was available in abundance. In the subsequent two decades there seems to have been little discussion of the comparative merits of rice and wheat. As had been true in other parts of Asia in the late 1940s, the main rice problem in China during this period was that there was too little of it.[70] Unlike in Japan and Taiwan, therefore, in mainland China the postwar decades did not usher in a wheat boom or general suspicion of rice.

Still, even in the People's Republic, improving rice eventually came to be seen as the indispensable step in improving nutritional status. This happened in the aftermath of the cataclysmic Cultural Revolution, as China reopened contact with Western countries and turned toward a market economy in the late 1970s. At that moment the crop scientist Yuan Longping, who had created high-yield, hybridized strains of rice, started to be hailed as a hero.[71] His innovation has been touted as China's contribution to the Green Revolution. As Sigrid Schmalzer has shown, such a reading obscures what was "Red" about Yuan's Maoist-era work. For example, a vast apparatus of collectivist science had supported it—Yuan was no lone innovator—and a socialist program to educate rural youth had produced Yuan's ablest research assistants.[72] Helpfully for the governing Communist Party, the tale also obscured the role that Mao's policies had played in depriving tens of millions of people of food. The glorification of Yuan's discovery implied that it hadn't been the wholesale disruption of the rural economy under collectivization, draconian controls on population movement, or unfair rationing that had left people undernourished. The technocratic narrative made the problem look simpler and more fundamental. It was that China had too many people, and existing strains of rice had yielded too little to feed them all—and Yuan's better rice could solve that. His triumphant story bolstered the regime's legitimacy rather than corroding it as the other explanation would have done.

In addition to making the ruling party look better, the post-Mao focus on improved rice reinforced nutritional imperialism. Once again, a problem of recent vintage, in this case nationwide famine and food shortages instigated by new policies, was made out to be a timeless feature of Chinese existence. Hadn't China always had insufficient food for its large population? And once again, science was presented as the solution. The story's ironic coda is that although a Chinese scientist had been the one to create the hybridized rice plant, before long it was owned by an American conglomerate. In 1979 Armand Hammer, the CEO of Occidental Petroleum, bought the right to grow Yuan's rice in the United States and to market it internationally.[73]

In other words, as much as perceptions of rice's health value in the early PRC differed from those in postwar Taiwan, by the 1980s rice had come to be seen even in mainland China as not quite good enough. As had been the case in Taiwan, this shift served both the government's own priorities—in Taiwan, to bring in foreign currency; in China, to preserve the regime's legitimacy—and American economic interests.

This chapter has explored how rice, central to many Chinese diets, came to seem like a food that made its eaters vulnerable to nutritional deficiencies and their consequences. The aspersions cast on rice did not originally come from scientific quarters. Instead, White observers with little experience of rice or of China linked their presumption that Chinese people ate rice almost exclusively with their presumption that Chinese people were passive, unimaginative, and easily dominated and created the epithet "rice eater" as shorthand for the whole bundle of stereotypes.

Nutrition scientists did not consciously traffic in such stereotypes, but their work on vitamin deficiencies nevertheless reinforced the message that Chinese people had been subsisting on an unfit staple. Vitamin deficiency diseases had recently become visible thanks to two kinds of reduction. The first was a reduction in diets themselves, which were tightly constrained in the modern institutional settings in which increasing numbers of people lived. The second was a reductive gaze of scientific inquiry, fixed on the chemical content of a plant or the weight of a rat. What emerged on the other side of experiments was a too-general, historically myopic conclusion: The Chinese diet, and particularly reliance on rice as a staple, cursed its eaters to vitamin insufficiency and all its consequences. Never mind that there was no such thing as a uniform, unchanging Chinese diet or that nationally visible vitamin-deficiency epidemics were a new phenomenon of the industrial and colonial age. A health crisis associated with modern economic and political change came to look like a longstanding defect of Chinese foodways.

Accordingly, to fix the problem reformers suggested changing both the problem grain and Chinese foodways. Improving rice and embracing wheat as a substitute could eliminate vitamin deficits and the shortcomings they were thought to provoke. Coincidentally, such changes stood to benefit Europeans and Americans financially by providing new markets for rice-altering technologies they created and for the wheat they produced in excess. Thus the circle of nutritional imperialism was complete: The expertise and resources of

European and American nations were seen as rescuing Chinese people from a nutritional plight that traditional diets had supposedly induced.

Even today, when the era of formal imperialism is more or less over, similar attitudes about rice can pop up. For example, a 2023 article in the British magazine *The Economist* titled "The Global Rice Crisis" associated rice with diabetes in Asia. Asian countries, the writer argued, needed to adopt better-engineered varieties of rice and better planting methods to "reduce environmental damage and increase yields." Their governments also needed to convince their people to eat less rice and more of other kinds of grains. Elsewhere, Chinese-language blog posts and chat threads proclaim white rice, with its high glycemic index, "the king of junk food"—a problem for Eastern civilization, one writer observes, since white rice as a staple food is a defining characteristic.[74] Such examples resonate strongly with the rhetoric of the early twentieth century that disparaged rice for being nutritionally unsound and endangering national progress. They blame disease associated with rice on rice itself, rather than on the socioeconomic factors that make eating a balanced diet difficult or impossible for many people.[75]

In the modern period, people have blamed rice for a lot. They have also invested rice with tremendous hope. But rice is both less and more than nutritional imperialism has made it out to be. It is *less* in the sense that it has never constituted the sole staple or the defining grain of Chinese civilization, and in the sense that tinkering with rice will not solve intractable public-health or geopolitical problems. It is also *more* in the sense that whatever its shortcomings when eaten alone, in combination with other foods rice has helped sustain vigorous populations for thousands of years. The epidemiological circumstances, political preoccupations, and scientific myopia of the modern period, which turned rice into a nutritionally suspect foodstuff, should not make us forget that.

In the next chapter we will turn our attention to a different problem that nutritional imperialists saw in twentieth-century Chinese diets: a lack of meat and other forms of animal protein. Popular literature sometimes linked this with the "rice problem," since a person who supposedly ate only rice counted as a type of vegetarian, but the science that this case implicates is different: protein rather than vitamin science. We will explore how squabbles about protein requirements reinforced the message that Chinese diets were deficient and possibly pathogenic.

3

The Wrong Kind of Protein

Vegetarianism and National Weakness

In 1927 the biochemist Hsien Wu (Wu Xian, 吳憲, 1893–1959) worried about his fellow Chinese citizens. The data showed that they were "small in stature" compared with American and English people and had "shorter lives with a higher mortality both among the adults and the infants." They seemed less able to resist infectious diseases such as trachoma and tuberculosis. They were also, he asserted, "over peaceful, non-persevering, non-progressive, non-enterprising," and "easily contented in the environment in which they find themselves."[1] What had caused this configuration of unwelcome traits? And what could be done about it? The causes were probably complex, Wu conceded, but he supposed that Chinese eating patterns had played a role. In particular, he suspected that overwhelmingly plant-based diets had helped make his compatriots sickly, passive, and uncompetitive. By eating more meat, milk, and eggs—like the Americans and the English—the Chinese could overcome their limitations.

Wu's credentials qualified him to pronounce judgment on Chinese diets. Some 15 years earlier he had been selected to study in the United States as a member of one of the earliest cohorts of Chinese students sent there with the support of a Boxer indemnity scholarship.[2] After earning an undergraduate degree at the Massachusetts Institute of Technology and a PhD studying protein at Harvard, he returned to China to work at the Rockefeller-founded Peking Union Medical College, one of the country's premier scientific institutions, where he taught physiological chemistry. By 1928 he had risen to become head of the biochemistry department and was beginning to train young scientists who would eventually number in the dozens and go on to lead research institutions across the country. He was well on his way to achieving the stature he retains posthumously today as one of the twentieth century's most influential Chinese scientists.

Wu published his critique in multiple outlets in English and in Chinese, and it attracted attention both at home and abroad.[3] His suggestion that a lack of animal protein had compromised Chinese health and character was nothing new, however. It merely lent precision and scientific gravitas to a claim that European and American observers had been making for decades: that China had been a "vegetarian nation" since ancient times and that this proclivity had shaped Chinese civilization. The commentators disagreed about whether it had shaped China for good or for ill, but they shared the belief that Chinese diets had long lacked meat and that that distinctive characteristic had made the Chinese people who they were.

This chapter examines claims about "vegetarian China." Observing how little meat, milk, and eggs Chinese people ate in the 1920s through 1940s compared with Europeans and Americans, Chinese nutrition scientists worried that their nation suffered from a collective protein deficiency. They were informed by the popular belief that eating meat conferred not only bulk and brawn but also drive and ingenuity. The New Nutritionists' conviction that suboptimal diet threatened nations as much as malnourishment also shaped their thinking. From their perspective, many generations of refusal or inability to eat meat had put the nation at a distinct disadvantage vis-à-vis Western imperial powers. They therefore saw increasing the animal protein in Chinese diets along European and American lines—the way the Japanese had done in recent decades—as a promising way to save China.

Such arguments bear the hallmarks of nutritional imperialism. As the previous chapter showed, early twentieth-century narratives exaggerated and homogenized the role of rice in Chinese diets across region and period; this chapter shows that they similarly ignored variety in Chinese consumption of meat. They inaccurately depicted early twentieth-century eating patterns as intrinsic to civilizations throughout history. They made China the foil of the West. And in some manifestations, as in Wu's assessment quoted above, they implied that Chinese political and economic subjection was in part a consequence of how Chinese diets fell short of European and American standards.

The Origins of China's Reputation as a "Vegetarian Nation"

It is revealing that when Wu published his attention-grabbing article in 1927, research into the composition of Chinese diets was in its infancy. The earliest survey had been conducted in 1923 by the American William H. Adolph and his students at Cheeloo University in Jinan, examining the diets of about 500 adults and children in North China.[4] The Chinese Physiological Society, in whose journal most such research would be published, was not founded until 1926. Wu himself conducted numerous studies comparing vegetarian with omnivorous rats, but he did most of this work after he had published his article. Even in the mid 1930s, as Wu's younger colleague Zheng Ji (Libin T. Cheng) later recounted, in China "there were only one or two dietary survey reports published," so Zheng felt impelled to conduct some of his own "in order to more completely grasp the true nutrition problems that existed among Chinese citizens." Even when he'd finished those surveys in the mid-1940s, Zheng lamented that "because of the limitations of time and manpower," the surveys "still fell far short of what I'd originally planned."[5]

Considering this relatively shallow pool of empirical evidence, one might reasonably ask why Wu felt so confident in declaring in 1927 that the "fundamental difference between the Chinese and the American diets . . . is that the former is mainly vegetarian while the latter contains a considerable proportion of animal food."[6] As it turns out, this view corresponded with popular perceptions among the global elite. Beginning in the nineteenth century, European and American observers had begun to refer to China as a "vegetarian nation." Just like the notion of "rice eaters" explored in chapter 2, this generalization both derived from and reinforced preexisting beliefs about the character of Chinese people and culture.

Along with wheat and dairy, meat—specifically, beef—was part of the trifecta that Rachel Laudan has identified as the core of Western power cuisine in the early twentieth century.[7] Scientists, doctors, and other scholars insisted that eating meat made societies progress. For example, in *Parcimony in Nutrition*, published in 1909, the English neurologist Sir James Crichton-Browne wrote, "It is clear that animal food has played a decisive part in human evolution, and that the craving for it has largely contributed to the advance of civilization. . . . All the more energetic races of the world and those most distinguished for intellectual capacity have been meat-eaters."[8]

Such a claim required a less-energetic, less-intelligent race of vegetarians to serve as a contrast, which political and economic philosophers such as the American Van Buren Denslow had earlier supplied. In 1888 he suggested that Chinese people had sunk over the course of their long history into a kind of passive torpor. Such people, he wrote, "will accept a vegetarian diet, will abhor innovation, and will shirk all contest, change, and revolution—hence will come under a bureaucratic, paternal despotism."[9] A doctor in Chicago even suggested that the Chinese, like the people of other "vegetarian" nations, had "small intestines two or three times as long as those of Americans and other flesh-eating nations." This implied that the two contrasting diets had provoked evolutionary changes resulting in something like different species of humans: Westerners, equipped to digest primarily meat in abundance, had shorter, wider intestines resembling those of carnivorous animals, while Chinese people had the kind of longer, narrower intestines seen in herbivores.

The convictions of these eminent authorities found blunter expression elsewhere, such as an advertisement reprinted widely between 1900 and 1905 across American newspapers—in the *Washington Post*, the *Atlanta Constitution*, the *St. Louis Post-Dispatch*, the *New York Tribune*, the *Detroit Free Press*, the *Cincinnati Enquirer*, the *Boston Daily Globe*, the *Hartford Courant*, the *Austin Statesman*, and the *Chicago Daily Tribune*, among others. The piece was written in the format of a news article, with a headline proclaiming "RULERS OF THE WORLD: Meat Eating Nations Are the Leaders in Every Branch of Human Achievement." The writer asserted, "The ruling nations of the world are meat eaters and history records that they always have been. . . . The Americans, English, French, Russians and Germans are meat eating nations and they are also the most energetic and most progressive." In contrast, "The weak races of people are the rice eating Chinese, Hindoos, and Siamese, regarded since the dawn of history as non-progressive, superstitious and inferior physically and mentally to the meat-eating nations who dominate them." That the ad was touting pills for meat eaters' indigestion may have made its bravado ring a little hollow—but the ad assured readers that eating meat upset their stomachs not because meat was "unwholesome" but because their modern urban lifestyles had brought on "disease or weakness."[10]

Such self-congratulatory celebrations of meat's civilization-building role revealed a marked ignorance of both China and history. The relative weakness and strength of peoples, whether measured in military, economic, or political terms, was far different in 1900 than it had been even a century prior. It may have been reasonable to describe northern European nations and the United States as "RULERS OF THE WORLD" in 1900, but not too long before, that title would have been contested or even laughable. The implication that the European elite had historically eaten more meat than the elite elsewhere also lacked a sound empirical basis, as did the claim that intestine lengths varied by race. And Denslow's suggestion that Chinese people eschewed revolution would surely have surprised the participants on both sides of the Taiping Rebellion (1850–1864), the deadliest uprising in human history, which had taken place when he was a young adult.

More specifically, the writers just cited were clearly unaware that China's heritage of carnivory was just as lavish as that of any other society. Textual and archaeological sources suggest that the elite in ancient China relished meat, which occupied an important place in their ritual and social life. As Roel Sterckx has shown, meat served as a medium for maintaining relationships: Rulers gave subordinates shares of their meat, and the subordinates, in turn, gave shares to others who were politically relevant to their lives.[11] A diverse range of species appeared on their tables, including both domesticated animals such as cows, horses, pigs, and chickens, and wild animals such as deer, rabbit, turtles, and a variety of birds and aquatic creatures. The historian Françoise Sabban reports that some of the earliest texts mention orangutan lips and elephant trunks as particular delicacies; bear paw was another. The philosopher Mencius famously asserted that just as he would choose righteousness over life, he would choose bear paw over fish—implying that just as a gourmand valued bar paw over plain old fish, a sage pursued righteousness even more determinedly than life itself.[12] Not only textual sources, such as poems and hunting statutes, but also archaeological evidence, such as the meat dishes preserved in aristocratic tombs, attest to the wide-ranging carnivory of the class that had access to animal flesh.[13] In fact, routine access to meat was such a notable status marker that *rou shi zhe* 肉食者, "meat eater," became a term for a government official.[14] For others—most people, that is—meat was available only a few times a year on special ritual occasions, such as during the lunar new year or following public sacrifices to gods or wandering ghosts.

Nor was there evidence that most Chinese people in the millennia following those ancient sources had forsworn meat eating, even if the amounts they ate in the early twentieth century were smaller than had by then become customary in the United States and Europe. Drawing on Guomindang government industry surveys from the 1930s, the historian Thomas David DuBois estimates that in Republican China each year more than 14 million cattle, 21 million pigs, and 61 million sheep were being slaughtered for food and concludes that "the scale and geography of animal slaughter indicate that meat was locally available" throughout the country.[15] Early Chinese nutrition surveys, though they were few and fragmentary, showed much the same thing: Most of the diets studied incorporated both meat and vegetable components. In fact, the diets that the scientists conducting the surveys described as "vegetarian" (*sushi xing* 素食型) were those that featured meat only once or twice a month—not instances of total abstinence.[16]

Still, it is not hard to see why some European and American observers came to the conclusion that meat avoidance was characteristic of Chinese culture. For one thing, meat abstinence was a prominent principle in Chinese Buddhism, one that distinguished it from Buddhism practiced elsewhere. Whereas early Buddhist texts from South Asia had been ambiguous about the moral status of meat eating, the Mahayana version of Buddhism that flourished in China took a much harder stance against it, buttressed by prohibitions from devout rulers such as the sixth-century-AD emperor Wu of Liang.[17] Eating meat, according to Chinese Buddhism, inured a person to slaughter and thus damaged their capacity for compassion, and this in turn limited their prospects for a good rebirth. Like alcohol or the five pungent foods (*wu xin* 五辛, such as garlic, leek and scallion), meat also was thought to cause excitement and stimulation and therefore to impede the kinds of activity that fostered enlightenment, such as solemn meditation. Later, animal welfare entered into consideration, but in a way that privileged concerns about human-human relationships above all. Since Buddhist doctrine held that living beings are reborn in different forms after death, one could never be sure that the pork or chicken one was eating was not the flesh of one's own deceased ancestor—a clear violation of filial piety.[18]

By the time Christian missionaries arrived in China to try to attract converts, avoiding meat had become such an established principle among committed Buddhists that getting them to eat it became part of the conversion process. The sixteenth-century Jesuit missionary Matteo Ricci argued that because God made animals for humans' benefit, it was "a form of ingratitude" to refuse to kill and eat them, as the scholar Eric Reinders explains.

Centuries later, Protestant missionaries tested the sincerity of Chinese converts from Buddhism by making them participate in a public meat meal, a kind of "baptism" to ritually break their decades-long or even lifetime abstention.[19]

Hsien Wu reinforced the narrative that meat avoidance was a characteristically Chinese problem and that Buddhism had helped make it so. Unlike Christian missionaries, Wu did not view vegetarianism as an obstacle to salvation; instead, he presented the advent of Buddhism in China as an ominous *nutritional* moment in China's history. "Superstitious disciples" of this religion, he explained, perceived killing living things as a sin. Consequently, Chinese people began to abstain from meat for long periods or for a lifetime, as opposed to the periodic short-term fasts that had been common in earlier ritual culture. And Buddhist abstention, he suggested, was "complete vegetarianism," which meant abstaining from all animal products, including fish, eggs, and dairy. Because of Buddhism, he contended, Chinese diets had become "lopsidedly vegetarian."[20]

Wu's conclusion overstates the impact of Buddhism, even the more rigorously vegetarian type, on Chinese diets as a whole. The historian Eric Greene has convincingly argued that Buddhist proselytizers adopted strict meat avoidance because it was a practice that distinguished them from the Chinese mainstream. Like Daoism, which appeared in the second century AD, Buddhism offered a kind of countercultural alternative to a state and scholarly culture that seemed to be in crisis. Both new faiths rejected the ancient "sacrificial bargain" that entailed offering slaughtered animals to the gods and ancestors in exchange for protection and assistance.[21] As Greene puts it, "Vegetarianism was one way Buddhists and Daoists could explain what and who they were *not*."[22] In other words, it was precisely *because* meat eating was such a prominent part of elite Chinese culture that avoiding meat became a signature characteristic of Chinese Buddhism. In India, where the Buddhists' rivals were themselves vegetarian ascetics, avoiding meat could not serve as a distinguishing marker. Thus, in contrast to their Chinese co-religionists, not only did Indian Buddhists not insist on vegetarianism, but some of their texts even presented vegetarianism as heretical or fanatical.

In effect the Chinese Buddhist clergy's contrarian practice of lifelong, absolute vegetarianism ironically highlights a broader cultural proclivity for meat *consumption*, not meat avoidance. The difficulty that meat-proscribing authorities faced in enforcing their dictates further supports this reading. When tracing the history of Buddhist vegetarianism, many scholars have pointed to the influence of the aforementioned Emperor Wu of Liang, himself a devout Buddhist and anti-meat, anti-alcohol zealot. In a famous document called *Discourse on Giving Up Alcohol and Meat* (*Duan jiu rou wen* 斷酒肉文, 518 AD), he prohibited the consumption of meat and alcohol by Buddhist monks and nuns. That he followed up with subsequent restatements of his prohibition suggests that it did not transform habits as promptly or completely as he had hoped—or as modern scholars have sometimes suggested—even among the clergy.[23] Similarly, a taboo against sacrificing cows and eating beef that emerged between the ninth and twelfth centuries AD seems to have been more honored in the breach than in the observance, judging from the way numerous sources felt compelled to insistently restate it.[24]

In sum, while Buddhism introduced powerful new rationales for meat abstention and helped inspire a sophisticated gastronomy of imitation meats, its overall impact on the protein content of Chinese diets was likely negligible.[25] Even if one believed, as many did in the early 1900s, that meat-light diets produced less fit people, the narrative that Buddhism had diverted Chinese people from an ancient, meat-rich heritage does not stand up to scrutiny.

In addition to China's distinctively vegetarian form of Buddhism, another thing that made some early twentieth-century observers see the country as perversely meat-averse was the contrast with Japan. Since the Meiji Restoration of 1868, Japanese officials had enthusiastically promoted meat eating. The Meiji emperor himself had famously eaten beef in 1872, signaling to his subjects the demise of a centuries-long taboo on beef consumption. Thereafter, officials and scholars encouraged Japanese people to eat more beef, both because they thought it would boost "health, growth and strength" and "intellectual and moral capacity," as Katarzyna Cwiertka has put it, and because some viewed it as prophylaxis against cholera.[26]

Their efforts highlight the intimate relationship between beef and imperialist expansion in East Asia. Aya Majima has shown that although the Japanese elite began extolling beef eating as a modern, civilized habit in the early 1870s, it was the colonization of Korea in 1910 that enabled many ordinary Japanese people to taste beef for the first time. Largely because meat was such an important part of Westerners' diets at a time when Westerners exercised tremendous influence over East Asian states, the Japanese elite associated it with greater strength and height. They were unable to boost supply from domestic sources as quickly as they wanted to, but once Japan took Korea as a colony, they systematized and expanded Korean beef farming and sent Korean cattle and beef to Japan. In addition, hundreds of thousands of Japanese people moved to Korea, where they could enjoy a more elite lifestyle as colonial expatriates, which included, for many of them, eating beef.[27]

Thus for Japan nutritional imperialism had an additional meaning. Not only did it inspire a dietary change along European and American lines, but also, realizing their dietary aspiration required building an empire. Western and Chinese observers of the time, though, usually posited the opposite causal relationship between Japan's meat eating and its growing empire, speculating that the former had conferred the physical prowess that made the latter possible.[28] In his 1927 article, Hsien Wu arranged the estimated daily meat consumption of the average citizen of 12 different countries in a table arranged from highest to lowest. In this table Japan ranked just above China at an estimated 25 grams of daily meat compared to China's estimated 20 (Table 3.1). Wu commented, "It would indeed be rash to say that all the

Table 3.1. Meat Consumption per Capita per Day

Country	Amount of Meat Consumed (per capita per day, in grams)
Australia	306
United States of America	149
Great Britain	130
France	92
Belgium and Holland	86
Austria-Hungary	79
Russia	59
Spain	61
Italy	29
Japan	25
China	20
Korea	5

Source: Based on Hsien Wu, "Chinese Diet in the Light of Modern Knowledge of Nutrition," *Chinese Social and Political Science Review* 11, no. 1 (January 1927): Table 6.

difference between China and Japan is due to the difference of five grams of meat consumed per capita per day by the two peoples. . . . But the general parallelism between the amount of animal foods consumed and the progressiveness of the people suggests some relation which is not accidental."[29]

As we have observed, empirical data on Chinese diets in the 1920s through 1940s was patchy. Wu's table does, however, highlight that some evidence did exist. For his table's China figure he drew on "data collected in Peking by the Morning Post for the month of April 1926." A survey conducted by a newspaper was probably not as sound as one conducted by trained social or biochemical scientists, but there is little reason to doubt its overall gist—that Chinese diets at the time included a far smaller share of meat than the diets of the richest and most industrialized nations. Even if the most extreme assertions of the contrast between so-called vegetarian nations and meat-eating ones were silly and misleading, I have seen no evidence to contradict the perception that in the aggregate, Chinese people ate less meat than Europeans and Americans in the early twentieth century. The historian Wilson Warren has estimated that the nations that consumed the most meat per capita by the end of the nineteenth century were Australia, New Zealand, the United States, Argentina, Britain, and Germany, in that order.[30] The difference that observers at the time perceived between Chinese and European (or European-descended) eaters' diets probably reflected reality.

Even so, this was not a longstanding cultural difference as observers took it to be. Relatively high levels of per-capita meat consumption across an entire population were a new phenomenon. Warren observes that meat eating in the late nineteenth century was strikingly different from what it had been earlier. It was associated with urbanization; slaughterhouses were largely in and around cities and were often supervised by municipal authorities. Starting in the 1880s refrigeration technology made the meat supply safer and made it possible to safely ship meat farther than before using refrigerated rail cars, refrigerated ships, refrigerated slaughterhouses, and refrigerated warehouses at transfer sites. Canned meat also became common. All of these changes put more meat in reach of more eaters, initially in the most industrialized and urbanized countries, with the result that meat became much more prominent in the diets of Europeans and people of European descent than it had ever been before.[31] But half a century earlier, the differences between these nations and others had not been so sharp.

Whether in Europe, the United States, or China, meat consumption had varied enormously between classes and regions within each nation.

So far, the story here is similar to the one about rice that we saw in chapter 2. In both cases, nineteenth- and twentieth-century observers identified a single feature of diverse diets as a marker of racial or cultural difference. In some instances, rice was the feature they focused on; in others it was meat. Chinese people became both "rice eaters" (never mind that many ate wheat-flour foods) and additionally a "vegetarian nation" (never mind that many ate meat). In addition to flattening contemporary diversity, vegetarian-nation arguments were teleological, presenting the eating habits of modern people as having crystallized in some ancient period and thus leading inexorably to the unequal distribution of power, wealth, and health in the present. By this reading, the powerful Europeans (and people of European descent) were timelessly meat eaters and the weaker Chinese a timelessly meat-free civilization.

The issue of meat eating differed from discussions of rice, however, because vegetarianism had a significant and committed constituency in late nineteenth-century Europe and the United States.[32] The term "vegetarian nation" as applied to China appeared most often *not* in jingoistic American newspapers, but in essays published by advocates of meatless eating. There, it expressed admiration. *Good Health*, the magazine of John Harvey Kellogg's Battle Creek Sanitarium, printed a number of articles between the 1880s and the first decade of the twentieth century praising the Chinese for their "vegetarian" diet, which the writers believed bore out the sanitarium's assurance that abstemious, meat-free eating promoted health and virtue. In 1903, the vegetarian activist William E. A. Axon published an article recounting "Some Chinese Arguments for Vegetarianism," an interesting pastiche of rationales from Chinese proverbs, early texts, an eighteenth-century scholar, and a popular tract against beef-eating written in the voice of an ox.[33]

Many admirers of China's presumed vegetarianism emphasized the strength and stamina of Chinese laborers. It is a theme apparent in some of their earliest tracts, such as the 1856 treatise that asserted that Chinese people "feed almost entirely on rice, confections, and fruits. . . . Those who are enabled to live well, and spend a temperate life, are possessed of great strength and agility."[34] By 1883, writers such as the Victorian vegetarian Howard Williams were asserting that Chinese workers were stronger than meat-eating Western workers. "In the East," he wrote, "there are at this day non-flesh eaters, who in feats of strength might put even our strongest men to the blush." Chinese laborers, he observed, were "notorious for their marvellous powers of endurance."[35] Similarly, an English civil servant who wrote a number of popular-science books devoted particular praise to "the common people of China" in his 1905 treatise advocating vegetarianism. Their diet had been "for ages derived almost entirely from plants." They were, he informed readers, smart and industrious, hardy, and exceptionally strong: "They amble along under loads that a strong Englishman could with difficulty raise from the ground." Citing "the Chinese Minister in Washington," he reported that Chinese laborers "are quite extraordinary in their capacity to bear heat and cold, and their energies are sustained by a few bowls of rice."[36]

Others praised the brilliance and creativity of the supposedly meatless Chinese. The author of a 1900 piece in the *Vegetarian Magazine*, the publication of the Vegetarian Society of America, defended the "vegetable eating nations" from the aspersions cast on them by a recent article. The offending essay had argued that a "high order of brain and nerve-power cannot come from a cheap, or limited, or vegetarian diet. . . . The vegetarian nations—India and China—invent nothing and lack thought and will power." In contrast, the pro-vegetarian rebuttal asserted that China had a rich literary tradition on par with the one the United States had inherited from Europe, and that its art "put to shame" American art. Moreover, its perceived deficits in military and other sciences reflected ethical constraints, according to this author, rather than any lack of aptitude. In short, they concluded, "It must be proved that vegetarianism causes lack of first-rate intelligence in some other way than by citing the Oriental as an example."[37]

The China enthusiasts were no better informed than the China detractors they battled. Judging from their biographies and the generality of most of their writings about China, writers of the pro-vegetarian tracts, like the meat boosters, did not have experience in China and were not personally acquainted with Chinese people. When they cited sources at all, the sources they cited were usually English-speaking travelers and residents in China—missionaries, diplomats, and businessmen. Visitors to the Battle Creek Sanitarium included more than a few recently returned or on hiatus from living in China, so *Good Health*'s contributors may have been able to consult such observers in person. But the vegetarians, like their opponents, viewed China as an abstraction that reinforced their own understanding of the meaning of meat. Neither side questioned the assumption that Chinese culture was by nature vegetarian or that the absence of meat had shaped Chinese character.

The one Chinese informant regularly cited in the English-language sources on Chinese vegetarianism shows how those sources belied the complex reality. This was Wu Tingfang (伍廷芳, 1842–1922), the "Chinese Minister in Washington" whom Williams had mentioned in his article. Wu twice served the Qing government as an ambassador to the United States. During his time in the United States he encountered *The Aristocracy of Health*, a 1904 book by the American Mary Henderson, which decried meat as one of humankind's "favorite poisons."[38] Following the book's advice, he gave up consuming meat as well as the other "poisons" Henderson identified, including alcohol, tobacco, tea, and coffee. The impact this new regimen seemed to have had on his health pleased Wu, and he became an energetic proponent of vegetarianism thereafter. He avidly read the work of American dietary reformers. He was most enthusiastic about Kellogg and visited Battle Creek more than once (Figure 3.1).[39]

He also promoted Chinese vegetarianism as an example for his American hosts. In his English-language memoir, *America through the Spectacles of an Oriental Diplomat*, Wu observed that "the United States has a Pure Food Department, but I think it might learn a great deal that it does not know if it were to send a commission to China to study life in the Buddhist monasteries, where only sanitary, healthful food is consumed."[40] He also recommended that his American readers try "an occasional, toothsome vegetarian repast as a set-off to the same round of fish, flesh, fowl and wine fumes" and boasted that "no people in the world can prepare such delicious vegetarian banquets as a Chinese culinary artist."[41] American journalists delighted in reporting his claim that his reformed diet would let him live to 200. (In the end, pneumonia claimed him at a respectable but not miraculous 79.)

Wu Tingfang's high-profile advocacy for meatless eating and his suggestion to an American audience that it was a particularly Chinese preference reinforced the "vegetarian nation" idea. But in truth it was American proponents of meatless eating who had converted Wu, and most of his vegetarian activism was aimed at his fellow countrymen. In the autumn of 1910, Wu founded the Cautious Diet and Health Society (*Shenshi weishenghui* 慎食衛生會) with fellow vegetarians Li Shizeng (李石曾) and Zhu Shaoping (朱少屏). Li, too, had

Figure 3.1 Wu Tingfang visiting Battle Creek Sanitarium, 1912. *Battle Creek Idea*, April 19, 1912, 4.

experienced a dietary epiphany abroad, in his case while living in Paris.[42] The society they established aimed to help people avoid illness and live longer by giving up "poisonous foods of blood and flesh" (*xue rou du pin* 血肉毒品).[43] They aspired to create tastier meat replacements, to establish a restaurant and central office building in Shanghai, and to produce informative publications about vegetarian living. Among their eminent proselytes were the leading academic Cai Yuanpei 蔡元培 and the well-known doctor and reformer Ding Fubao 丁福保.

In contrast to Wu's English-language publications, in which he had praised Chinese—and particularly Buddhist—contributions to vegetarian cuisine, the Cautious Diet and Health Society identified vegetarianism as a new, modern, and *Western* practice. They wrote that China's long "nourishing life" (*yangsheng* 養生) tradition had emphasized eating bland and pure foods, which could achieve some of the same effects, but argued that *yangsheng* teachings had not zeroed in on the important distinction between plant and animal foods. That was what modern vegetarians were doing. As the historian Angela Leung has noted, for his Chinese audiences Wu tended to downplay Buddhist tradition and emphasize the scientific foundation of his advice.[44]

In sum, by the early twentieth century, China was perceived internationally as a vegetarian nation. American and European writers showered this supposedly Chinese trait with either praise or derision, depending on their own orientation toward meat eating, but seldom questioned whether the characterization was accurate. The Chinese elite, meanwhile, were well aware that the society they inhabited was not a meatless one. Prominent figures, in some cases inspired by Chinese life-nourishing tradition, in some cases by Buddhism, and in some by vegetarian movements abroad, sought to change that. Like their American and European counterparts, Chinese vegetarian advocates were a vocal minority in their own society. By the 1930s, however, the voices of professional nutrition scientists had begun to drown out the amateurs, including the vegetarians. On the whole, the scientists concurred that China was too vegetarian for its own health and progress.

Lending "Scientific Dignity to a Popular Belief"

We have seen that Hsien Wu, the most prominent figure in the first generation of Chinese nutrition scientists, intuited as early as 1927 that the vegetarian bent of Chinese diets had affected the nation's health and character. The intuition was grounded partly in popular views that eating meat conferred strength and ambition and partly in a dubious reading of China's dietary history. But Wu was a scientist, not a pundit or a historian; having ventured a hypothesis, his next task was to test it experimentally. Starting in the late 1920s, therefore, Wu and his colleagues fed regimens meant to imitate Chinese diets to white rats meant to serve as a proxy for the Chinese people. Though the data varied, the scientists concluded that Chinese people's limited meat consumption had left them deficient—suffering not, as in the case of rice, from a vitamin deficiency, but instead, or in addition, from a protein deficiency. Since early twentieth-century developments in protein science helped pathologize Chinese vegetarianism, it is worth surveying the development of that science here.

Like the concept of vitamins, the concept of protein is vague, encompassing a wide range of chemical compounds. Just as the word "vitamins," meaning "a vital substance of unknown chemical composition," communicates the indispensability of this class of substances without specifying what, exactly, they are, "protein" communicates importance but not identity. The term derives from a Greek word meaning "primary" or "first," so it denotes simply a substance of the first importance in foods. Like Casimir Funk, who coined the word "vitamine," the Swedish chemist who coined the word "protein" originally believed that everything that fit under this new rubric shared a chemical structure.[45] Both were mistaken—later experiments proved that not all "vitamines" contained the amine group and not all proteins contained the so-called "protein radical." Nevertheless, the categories persisted.

Proteins are big, complex molecules that combine amino acids. The precise combination varies widely and differs across species and in different organs of a single animal. Proteins are found in both plant- and animal-derived foods, though they account for a larger proportion of animal-derived foods than of plant-derived ones. They are a class of thing, not a specific chemical compound. Protein, in other words, is protean. Nor is there a single standard way to measure the protein content of a food. The most direct method is to assess a sample's amino acid content, but that is expensive, so more indirect methods, such as measuring nitrogen content as a proxy, are generally used.[46]

Although the meaning of "protein" has changed a number of times since the term was coined, protein has never lost its status as a central—and perhaps the *most* essential—macronutrient. The nineteenth-century chemist Justus von Liebig maintained that protein was "the only true nutrient."[47] In published tabulations of the nutritional content of foods, protein was often listed first, before carbohydrates or fats or anything else, reflecting its presumed supremacy. Even today, protein remains "the last macronutrient standing," as the food journalist Bee Wilson has put it. One seldom hears concerns about whether or not people are getting enough fats or carbohydrates in their diets, but protein is nearly an obsession.[48]

Although it has become as ardent a fixation of Chinese eaters as of eaters elsewhere in the world, protein's career in China began a bit later than in Europe and North America. In contrast to the English word "protein," whose derivation from the Greek "first" emphasizes its primacy as a nutrient, the Chinese version of the term evokes a particular food: chicken eggs. *Danbaizhi* 蛋白質 literally means "egg-white matter." As is true of many technical terms in modern Chinese, this one was created in Japan, and that, in turn, was a translation of the Dutch *eiwit*, drawn from the German *Eiweiß*, "albumen" or "egg white."[49] It is thus concrete where "protein" is abstract, and it implies that eggs epitomize the category.

Danbaizhi appeared occasionally in newspaper advertisements in the 1900s, but it was in the 1910s that Chinese readers began to encounter the term more frequently. It does not, at first, seem to have been an overriding concern. When the Cautious Diet and Health Society was founded in 1910, for instance, its members were not especially concerned about protein. While they recognized it as important, in their eyes it was no more important than other macronutrients. In fact, Li Shizeng argued that vegetables were *more* complete than meats in their nutritional profile, containing not only protein but also starch, fats, and minerals. Most of meat's nutritional value, by contrast, came from its protein content, making it a less-whole source of sustenance in his view. Meat, moreover, contained "poisonous matter" (*du zhi* 毒質) that the body needed somehow to overcome.[50] The idea that meat was poisonous resonated with the still relatively new knowledge of germs and also reflected widespread experience with tainted meats—reports of trichinosis and of bovine tuberculosis were common even in articles with no vegetarian agenda.

Articles about dietetics in the magazines and newspapers of the 1910s similarly said little about meat's protein content, instead presenting plant foods as both less dangerous and more nutritious. The anarchist group called the Huiming 晦鳴 (literally "Calling in the Darkness") Society declared in 1913, "Meat contains a lot of poisonous matter and microbes that are the root of all brain, muscle, intestine, stomach, skin, and circulatory diseases. But plants don't have these things." Furthermore, they continued, "The nutrients in soybeans are richer than those in meat."[51] The following year a piece in a general interest magazine asserted, "Doctors today have tested what's in meat, like pork, fish, chicken, duck, and so forth and they all have a lot of poisonous material. Polluted blood produces all kinds of diseases. I have also heard that it has the seeds of infection and is a distinctive disease medium."[52] By contrast, plant foods were safe and wholesome.

By the mid-1920s, however, China's earliest practitioners of nutrition science had entered the conversation, and they put a premium on protein. An implicit question was whether Chinese eaters got enough of it. This proved tough to assess, in part because there was considerable international scientific debate about what the baseline ought to be. How much protein should a normal diet contain? Wilbur Atwater, the seminal figure in American nutrition research, had accepted and propagated the protein and energy standards proposed by the German physiologist Carl von Voit before him, recommending 100 grams of protein per day for an adult man—or more, depending on the intensity of his daily work. This consensus remained unchallenged until the work of the American Russell Chittenden began to attract attention in the first decade of the 1900s.

Chittenden, a Yale chemist with an affinity for the contrarian dietary movements of his day, believed the 100-gram figure was much too high.[53] To test the effect of a lower-protein diet, Chittenden first experimented on himself, subsisting on less than half of that standard. Suffering no ill consequences, in 1904–1905 he undertook more-systematic studies in which colleagues, soldiers, and athletes ate diets with only about 50 or 60 grams of protein. Even after five or six months the experimental subjects seemed to be maintaining their physical fitness and some even reported feeling greater well-being. Therefore, Chittenden concluded, one could safely consume far less protein than his colleagues had been insisting on. Indeed, it might even benefit a person's health to do so.

The quality of Chittenden's research and his prestigious academic pedigree made his arguments difficult to dismiss. The debate that he had stirred up featured in the international conferences hosted by the British Medical Association in 1906 and 1911. According to the notes from the 1906 association meeting, Chittenden's session on "Over-Nutrition and Under-Nutrition, with Special Reference to Proteid Metabolism" was well attended, and the attendees mostly praised his experiments. Some expressed reservations, though. For example, the well-known British physiologist W. D. Halliburton brought up the "low nitrogenous intake in vegetarian nations" and asserted the conventional view that "such people showed less resistance to privation and to bacterial invasions than meat-eating people." He pointed out, moreover, that some observers had attributed the "recent vigor and success" of the Japanese—who had a year earlier scored a surprising win over Russia in war—to "their tendency to give up vegetarianism and take larger quantities of animal food." Chittenden's data, he implied, failed to account for such phenomena. In reply the Harvard chemist Otto Folin, a mentor of and close collaborator with Hsien Wu, suggested that beliefs about "vegetarian nations" were unfounded. He argued that the main reason the high-protein Voit standard remained in place, despite the evidence accumulating against it, was because of a vague "conviction that the more effective portions of mankind indulged in a diet of this kind." Voit's experiments, in his view, had not revealed a natural law; instead, they had merely "lent scientific dignity to a popular belief."[54]

Indeed, it is easy to find proponents of the Voit standard articulating some version of that popular belief. David McCay, a British medical officer and professor of physiology at the Medical College of Calcutta, published a book in 1912 called *The Protein Element in Nutrition* attacking Chittenden's views. After exploring the question of protein from a number of different angles, he arrived at the final chapter of the book, clearly meant to be the coup de grâce for Chittenden's low-protein standard. The chapter, called "THE EFFECTS OF THE LEVEL OF PROTEIN METABOLISM ON THE PHYSIQUE AND GENERAL EFFICIENCY OF DIFFERENT TROPICAL TRIBES AND RACES," drew on beliefs common among colonial officers about the relative fitness and manliness of different groups of Britain's Indian subjects. He contrasted the supposedly effeminate, lazy Bengali with the "tribes which supply recruits for the Indian army," which he saw as "superior in physique, courage, boldness, and generally manly vigour."[55] Manliness, vigor, and energy—those unquantifiable yet, in McCay's view, easily comparable qualities—were the main effect of protein-rich eating. As he put it, "The higher the level of protein interchange, the more robust and energetic, and the more manly the race."[56]

Ultimately, then, while McCay's book considered many scientific experiments and deployed many facts and figures, he approached the protein question with the same assumption as the advertisement for indigestion pills that we saw earlier. He wrote, "All the successful races have habitually consumed protein far in excess of the amount Chittenden believes sufficient, whereas those who have adhered to a low protein standard have not progressed physically, mentally, or morally. The poorer races of India and the tropics illustrate this truth, and show that the conquering races and nations are all high protein feeders."[57] In other words, Meat Eating Nations Are the Leaders in Every Branch of Human Achievement.

Chinese scientists were no less concerned about protein than their European and American counterparts. Hsien Wu was especially attuned to the stuff; he was one of the first researchers in the world to come up with a theory about what causes proteins to lose their molecular structure.[58] He immersed himself in the protein debates. In 1926 he published an article in the Chinese journal *Science* (*Kexue* 科學) that introduced then-current knowledge about protein. After offering a definition ("an important constituent of cells" was as specific as he got), he offered a kind of blow-by-blow description of changes in scientists' understanding, beginning with Voit and Atwater and dwelling at length on Chittenden.

He explained Chittenden's view that eating too much protein harmed the body by generating damaging waste products, overworking the liver and kidneys, and stimulating the metabolism in a way that disrupted the whole body's economy.[59] This view resonated with what Li Shizeng and others were writing in the popular press about the poisons that meat contained and generated. It accorded, too, with a longstanding Daoist belief that overfeeding could shorten lifespan—in the Daoist framing, this was because pernicious beings lived in a person's gut and fed off of the food they ate, sustaining themselves while shortening their host's life.[60] Wu credited Chittenden for his innovative experiments, conceding that they "definitely had value for the physiological science of that era" some 10 years earlier.

Nevertheless, Wu rejected Chittenden's lower standard, arguing that his nine-month-long experiments were too brief to convince. "What's appropriate for a short-term diet isn't necessarily appropriate for a lifetime diet," he pointed out. "In other words, although eating a low-protein diet for nine months might not cause any harm, following it for years might produce obstacles to health." And so, he concluded, 75–115 grams of protein per day remained the safest guideline for an adult male engaged in moderate labor.[61]

Moving forward, then, Chinese scientists sought to determine whether diets in their country met this higher Voit-Atwater-inspired standard. The historian Jia-Chen Fu has examined in detail what they found. The data the scientists drew on—biometric and dietary surveys from different parts of the country, measurements of the protein content of common Chinese foods, and especially feeding experiments with rats—were messy and often contradicted one another or challenged researchers' assumptions. But the conclusions they drew, as is often true of scientific reports, tidied up the mess and overlooked the complications.

For example, researchers assumed that Chinese people were shorter than Westerners. While some comparisons bore out that assumption, there were also examples that showed the opposite; one Lester Institute scientist observed that the several hundred Cantonese boys he'd measured were taller on average than same-age British boys reported in another study. Similarly, scientists expected the rats they raised on vegetarian diets to be smaller, shorter lived, and less reproductively successful than those raised on mixed plant-and-meat diets, and they generally were—except when they weren't. And if the researchers surveying diets in Shanghai, Beijing, and Shandong expected the protein content of those diets to fall short of established standards, they must have been surprised: Their studies consistently showed that the families surveyed were consuming as much protein as international standards recommended.[62]

Such surprises weren't analyzed or much remarked on. The contradictory data about heights and rats could reasonably be discounted as outliers, since most of the studies did yield the expected results (i.e., southern Chinese bodies that were shorter than northern Chinese or Western bodies, and vegetarian rats that were scrawnier than their omnivore relatives). The perfectly adequate protein levels of observed Chinese diets, however, could not be explained away so easily, because the results were quite consistent. The diets of surveyed populations reliably fell in Hsien Wu's sweet spot of 75–115 grams of protein per day *despite* their negligible amounts of meat and other animal-derived products. They drew 80–95 grams of protein almost exclusively from grains and vegetables.

Did this mean that, contrary to expectation, Chinese diets provided adequate protein? It did not, according to the scientists. Research into the chemical constitution of different proteins had recently revealed that they were not all created equal. Proteins were made up of amino acids; it was unclear at the time exactly how many amino acids there were in total, but most scientists thought somewhere between 16 and 20, and each kind of protein had its own unique mixture of them. To build and maintain a human body, one needed to consume the kinds of amino acids that comprised it. A protein derived from an animal source contained a larger number of the types of amino acids found in the human body than did a protein derived from a plant source, and therefore animal protein's "physiological value" or "biological value" was higher.

Maximizing biological value, Hsien Wu explained in his *Kexue* article, meant trying to match the amino-acid profile of the protein ingested as closely as possible to that of the eater's flesh. He wrote, "Theoretically speaking, if you use horse protein to feed horses, and sheep protein to feed sheep, then the [feed's biological] value should be 100%. You should use a similar protein as food, and thus the composition of the amino acids should suit the needs [of the animal] and be neither excessive nor insufficient." This modern discovery, he declared, validated the Chinese-medicine practice of using animal livers to strengthen an ailing liver, and animal brains to tonify the brain.[63] He did not draw the obvious conclusion that, by the same reasoning, human flesh must offer a human eater the highest-value protein of all and that cannibalism was likewise "deeply compatible with scientific principles." Perhaps sensing the uncomfortable implication of his similar-protein prescription, however, he clarified that animals need not eat others of their own species, but that when feeding them one should choose proteins from living beings whose "animal systems" resembled the animal's own. Thus, "If you are feeding horses, donkey meat is better than chicken meat. If you are feeding geese, chicken meat is better than donkey meat." As for plants, their proteins were even less useful for humans than a chicken's were for a horse or a donkey's for a goose.[64] Plant-derived proteins were, moreover, more difficult to digest than animal-derived proteins.

The upshot of all of this was that even if the gross amount of protein that Chinese eaters consumed met an international standard, "The protein in Chinese people's diets is inferior to the protein in Westerners' diets."[65] So while a diet that supplied 80–95 grams' worth of protein per day might, on the surface, look nutritionally adequate, if that protein was mostly coming from plants then the amount of *usable* protein was likely far less, perhaps as little as 65 grams. Therefore, "What is a moderate allowance in the American diet becomes a bare minimum in the Chinese diet. Doubtless the German and American physiologists had in mind the proteins of milk and cheese, beef and chicken, and not those of wheat and rice, millet and kaoliang, when they proposed their standards."[66] Wu suspected that Chinese diets, though protein-sufficient on the surface, hid a deeper deficiency. Even Chinese people who appeared to be robust might, therefore, actually be suffering from a deficit. The manual laborers whose stamina had invoked admiration and anxiety around the world might in fact be malnourished; given a proper diet with meat, milk and eggs, they could have been even stronger, he thought, and certainly bigger.[67]

Size mattered to Wu—a lot. In the experiments he and his colleagues conducted comparing omnivorous with vegetarian rats, they examined many outcomes: How long did they live? How many offspring did they have, and did their offspring survive? How many developed cataracts? Were their calcium levels adequate? And so forth.[68] But the main index of health and success, and the one that most consistently showed the omnivores at an advantage, was how fat they got and how quickly. In his *Introduction to Nutrition* (*Yingyang gailun* 營養概論), Wu explained one such experiment. "I once took a number of rats from the same litter and separated them into two different cages. I fed one cage a Chinese-like diet, and the other cage an American-like diet." All the rats were able to grow up and reproduce and had no apparent disease, but the ones on the "Chinese diet" weighed less at every age than the ones on the "American diet." He illustrated the point with a series of photographs of two rats lined up side by side at different stages of development, a plump one labeled "rat who ate the American diet" next to a smaller, skinnier one labeled "rat who ate the Chinese diet" (Figure 3.2).[69] Although other indices were less consistent—for example, in one study the female vegetarian rats long outlived their omnivorous sisters—a mixed diet seemed generally to confer an advantage in length and weight.

As his textbook makes apparent, Wu intended experiments like this one to undermine racism, not reinforce it. The mini-experiment explained above supported his argument that a commonly remarked difference between Chinese people and Westerners—that the former were on average shorter than the latter—was not an innate racial characteristic. It was more likely to reflect differences in environment, of which nutrition was an important element. While inherited characteristics might be immutable, or at least tough to change, one could reform a diet immediately and see results even in the present generation. How long might it take to reverse the subtler consequences of centuries of meatless diets? On that Wu was less clear; presumably it was harder to assess how "over peaceful" or "non-

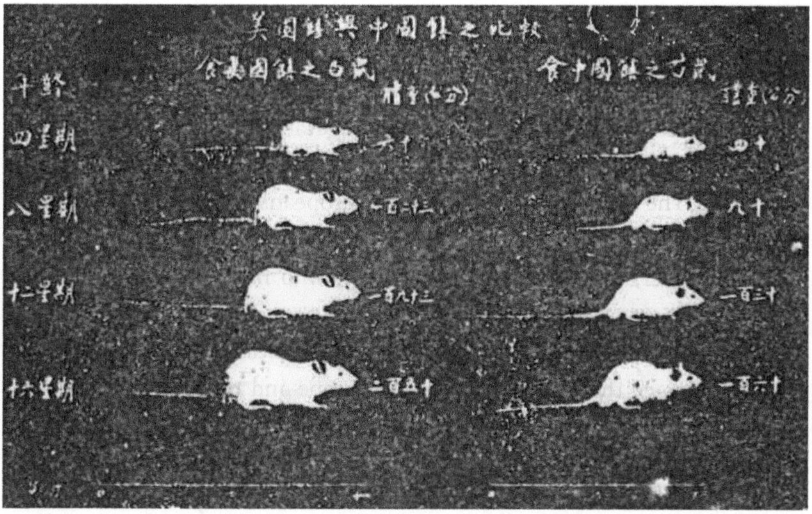

Figure 3.2 A comparison of rats fed the American diet and the Chinese diet. Rats fed the "American diet" are on the left, and rats fed the "Chinese diet" on the right. From Wu Xian, *Yingyang gailun* 營養概論 [Introduction to nutrition] (Taipei: Taiwan Commercial Press, 1967), 141.

enterprising" the different groups of rats were. Height and heft, however, could be rapidly improved.

In his effort to dismantle the essentialist view that Chinese people were shorter and smaller than others by nature, Wu reinforced a different essentialism, the nutritional-imperialist idea that there was a uniform Chinese diet and that it was inferior to the American one. Thus, reforming their diets along American lines was a sure way for Chinese people to improve themselves.

The point here is not to argue that Chinese citizens in the 1920s to 1940s were on average as well-nourished and healthy as citizens of northern Europe and North America. This was an era of endemic conflict and poverty in China. Although our data from the time are thin, it would be surprising, given the context, if poor nourishment had *not* been a problem for people living through it.[70] But this malnutrition was a social and economic problem, not a matter of obstinately sticking to a backward diet inherited from the ancient past. Similarly, we must concede that adding meat, milk, and eggs to more Chinese diets in this period would likely have improved people's nutritional status, health, and well-being. But that reflects the way these calorie-dense foods would have worked against the effects of hunger more than the protein boost provided by animal sources.

To summarize, well before the advent of nutrition science, Western observers had come to think of China as a "vegetarian nation," and many had begun to associate a lack of meat in Chinese diets with the country's political impotence. The simple narrative was that Europeans, made big, strong, ambitious, and ingenious by their long custom of eating meat, had prevailed over smaller, weaker, more complacent Chinese, who had long ago abandoned meat eating because of ideology. The reality was more complicated. While it was true in the late nineteenth century that Chinese eaters consumed less meat on average than those in Western Europe and the United States, this reflected a recent increase in the amount of meat that Europeans and Americans were eating, rather than an ancient dietary dichotomy. Moreover, while the comparative abundance of animal-derived foods may indeed have contributed to increasing height, weight, and longevity among Euro-Americans in this period, it was only part of a more general improvement in nutrition—one that was in part facilitated by colonial conquests that gave people in the metropole access to more calories and more diverse foods than before. The simple narrative distorted history in another way, too: It obscured the complex, contextual approaches to meat in Chinese political, social, and medical traditions—approaches that had structured relationships between meat and health for many centuries before nutrition science emerged.

In the 1920s Hsien Wu, the most prominent Chinese nutrition scientist of his generation, implicitly accepted the popular narrative, but he gave it a scientific veneer and framed it in more specific terms. Chinese people, he suggested, suffered from a protein deficit. Though surveys showed that the amount of protein in their daily diets fell within the range recommended by experts, because it was plant- rather than animal-derived it was probably still inadequate. Wu insinuated that any number of defects, from smallness to weak immunity to "non-progressiveness" might stem from this deficiency, but no one had yet proven a direct connection. Thus many generations of white rats served as a proxy for the Chinese nation in the experiments of the 1930s, eating their way through monotonous diets labeled "Chinese" or "vegetarian" in order to demonstrate the kinds of population-level shortcomings that might ensue from an unwise combination of foods. What they most consistently demonstrated was smallness and lightness. Since those were among the physical characteristics that emblematized poor health and inferiority, these experiments seemed to validate the popular view of China as a nation hobbled by a lack of meat.

The Global "Protein Gap"

China was not the only place where this form of nutritional imperialism materialized. In the first half of the twentieth century, scientists in other poor countries and in colonies frequently asked whether diets there contained enough animal-derived protein and frequently concluded that they did not. This belief persisted and became an article of faith in the era of decolonization, enshrined in the concept of a "protein gap" between the Third World and the First. While it was originally based on observations about kwashiorkor, a severe form of malnutrition first observed in Ghana in the 1930s, by the mid-1950s the World Health Organization (WHO) and the Food and Agriculture Organization (FAO) had decided that "protein malnutrition" was the number-one nutrition problem facing the world. To tackle this problem the US National Research Council created a Committee on Protein Malnutrition, and WHO established a Protein Advisory Group. A member of the UN's Committee on Nutrition declared in 1961 that international experts had "moved from the era of vitamin research to protein research."[71] Enterprising companies in the rich world began producing protein-rich supplements from their countries' agricultural surpluses and industrial waste—fish flour, algae burgers, corn-soy milk, wheat-soy blend—intending to sell them cheaply to the protein-deficient masses elsewhere.[72] The "protein gap," like the "rice problem" examined in the previous chapter, thus became a way to both unload a rich country's excess and burnish its moral profile.

In the midst of all this frenetic activity, few acknowledged a basic problem: Namely, there was no good evidence that a lack of protein actually *was* a main cause of malnourishment in poor countries. But the belief accorded so well with preexisting assumptions that the composition of Euro-American diets was exemplary and that rich nations were saving so-called underdeveloped peoples from their own ignorance and poverty that it persisted for decades.

By the early 1970s, the idea of a global protein gap had become unsustainable. The large-scale production and provision of protein-rich supplements had made no dent in global malnutrition, and evidence was piling up that kwashiorkor was exceptional, not representative. Most undernourished people suffered from a calorie deficit—in other words, from diffuse problems caused by not getting enough to eat—rather than a protein deficit. One could not solve that bigger problem by crafting bespoke supplements; only when people in poor societies had better access to more food and more diverse foods would the underlying problem begin to ease. Besides, in the mid-1960s the FAO and WHO had radically reduced the amount of protein that they recommended the average human consume daily, bringing it more in line with what Chittenden had said was necessary back at the turn of the century. As one nutritionist observed, this closed the imagined protein gap in a single bureaucratic stroke.[73]

One of the most visible expressions of scientists' growing discontent with the protein-deficiency paradigm was an article titled "The Great Protein Fiasco" published in 1974 in the *Lancet*, perhaps the most prestigious medical journal in the world. In it, the nutrition scientist Donald McLaren decried the waste of money, time, and children's lives that the scientific community's fixation on protein had occasioned.[74] Within a few years, recognizing the validity of the critics' arguments, WHO's Protein Advisory Group quietly disbanded, and scientists and policymakers ceased talking about a protein gap or protein crisis. So understated was the climbdown, in contrast to the earlier hyperbole calling protein deficiency "the most serious and widespread problem in the world," that few people outside of nutrition science even noticed it.[75]

Had Hsien Wu lived long enough to witness the fall of the protein-gap idea, perhaps he would have reconsidered how much the different amino-acid profiles of the proteins that Chinese and American people ate could account for differences in national character or political power. But he died of coronary thrombosis in 1959, having moved to the United States 12 years earlier. At the time, nutritionists' protein fixation was at its zenith and the United States—coincidentally or not—was nearing its peak rates of heart disease, which have declined continuously since the mid-1960s.[76]

In the remaining chapters we will examine two newer examples of nutritional imperialism: the invention of lactose intolerance and "Asian Flush." The prejudices underlying these two concepts can be traced to the same late nineteenth- and early twentieth-century context that inspired the ideas of Oriental metabolism, rice sickness, and meat deficiency. The science implicated, however, is more recent than what we have seen so far. Not only, it seems, are we living with the vestiges of older episodes of pathologization; we are also seeing new imagined pathologies develop.

4

The White Milk Burden

The Invention of Lactose Intolerance

In 1914 Berthold Laufer, a specialist in East Asian languages who was born and raised in Germany and made his career in the United States, ruminated about Chinese people's relationship to animal milk, which he saw as peculiar. In an essay published in the *Journal of Race Development*, he declared that "the entire East-Asiatic world, inclusive of China, Korea, Japan, Indo-China, and all Malayans, does not take animal milk for food, and evinces a deep-rooted aversion toward it; and this was the state of affairs even in remotest times." By contrast, Indo-European peoples "are all milk-drinkers, and were so in early historical times."[1] In Laufer's view this showed that the Chinese were stuck at a preliminary stage of agricultural development. They used cattle for plowing but not for milking, whereas Europeans "subsequently advanced to a further stage of development, from which the Chinese were debarred,—the acquisition of [*sic*] dairy economy."[2] Speculating about the effects of this divergence, Laufer idly remarked that it might bear on "mental achievements"—specifically, the creation of epic poetry. He thought it a "curious coincidence" that the Indo-Europeans who had produced the world's "epic songs are also milk-consumers, while those abstaining from milk are deficient in epic poems."[3]

Attributing *Beowulf* to milk drinking might seem silly to us now, but Laufer was not alone in viewing dairy deficiency as a signature of Chinese civilization. For him the main consequence was literary, but his contemporaries blamed missing milk for all sorts of other unfortunate characteristics, from the political to the physical to the moral. Milk thus joined the litany of supposedly Chinese deficiencies that inspired research questions in nutrition science, including the three we have explored in earlier chapters.

When it came to milk, observers wondered whether an inherited inability to metabolize it, which scientists called "lactose intolerance," had prevented Chinese people from developing what they saw as a proper dairy culture—that is, one that resembled the large-scale milk industries that began in European countries and the United States in the modern period.

This chapter explores how modernizers erased from view a Chinese milk tradition, succeeded in building a large-scale dairy industry on the model of the United States, and pathologized Chinese bodies as defective when they struggled to process prodigious amounts of fresh milk. This exemplifies nutritional imperialism as clearly as our previous topics did, but there is a difference. Whereas beliefs about "Oriental" metabolism, rice-related debility, and China as a "vegetarian nation" all peaked by the middle of the twentieth century, lactose intolerance was an invention of the 1960s and 1970s, and since then it has become embedded in scientific research programs and commercial development. It thus demonstrates that nutritional imperialism is not a thing of the past but a persistent inclination that continues to pop up in new places.

Although a number of histories of fresh milk's rise and entrenchment in modern societies and economies has appeared in recent years, most of these histories have been confined to a Euro-American perspective.[4] When they mention lactose intolerance, it is as a kind of aside, a shorthand explanation for why milk did not matter as much in most other parts of the world.[5] At the same time, archaeologists and paleoanthropologists have grown curious about lactose *tolerance*, a genetic trait that spread through a few small human populations 6,000 to 8,000 years ago.[6] Their studies focus on identifying the origins of lactose tolerance and explaining its spread. While this sort of genetic and anthropological research is interesting, it has unintentionally reinforced nutritional imperialism by elevating a quirk of metabolism into a core element of historical identity. Like other scholars, I argue that we should not accept at face value modern claims that healthy diets require dairy. When it comes to lactose intolerance, though, most scholars have accepted that it is a condition characteristic of Chinese people that justifies an aversion to dairy and strengthens the case against propagating milk culture there. Though I once shared this perspective, I have since come to see lactose intolerance itself as an insidious concept that reinforces erroneous beliefs like Laufer's in two tiers of civilization, those with milk and those without.[7]

Rethinking Lactophobia in Traditional China

As we saw in the previous chapter, Hsien Wu (Wu Xian 吳憲), the father of Chinese nutrition science, believed that Chinese people had undervalued and sometimes even avoided meat from a very early point in their history. When it came to milk, Wu's 1928 *Introduction to Nutrition* (*Yingyang gailun* 營養概論) offered a similar take, this time explicitly framed as a comparison with European countries. Wu asserted, "There is one way in which European diets differ from ours that bears paying close attention to: Europeans consume milk, and Chinese people do not consume milk. You can see milk used as food in the Christian Bible. It probably began as a habit of the nomads of central Asia and was adopted by Europeans." By contrast, he claimed, "In my country's classics there is no record of consuming dairy. In recent years it has occasionally been consumed; that is the result of European culture flowing to the East."[8]

From Wu's perspective and that of virtually all of his fellow nutrition scientists in the first half of the twentieth century—both Chinese and international—milk was the *sine qua non* of good diet. Chinese people, they believed, had suffered from a virtually complete absence of the precious stuff from the earliest days of their civilization until the entrance of Western ideas and practices in the nineteenth century.

Many modern scholars have agreed with this narrative, suggesting that milk was peripheral to or absent from the Chinese diet for most of history. The anthropologist Frederick J. Simoons, who divided the globe into "milking" and "non-milking" cultures, saw the Chinese as an example of a non-milking people; others have written about China as a historically "lactophobic" society in which milk was the object of revulsion.[9] Some presented the relative scarcity of milk in traditional Chinese diets as a puzzle. In 2002 the question "Why did the Chinese fail to incorporate milk and milk products into their food system?" was common enough that the biochemist and historian H. T. Huang felt compelled to write an article responding to it.[10]

Others supplied answers to this supposed puzzle by pointing to cultural or physiological factors that worked against dairy's assimilation in China. E. N. Anderson and Marja L. Anderson, for example, suggested that the association of dairy with koumiss-drinking Mongols or yak butter–eating Tibetans—peoples the Han Chinese considered less civilized than themselves—made it less palatable for late-imperial Chinese.[11] In addition to cultural prejudices, they noted, there was a physiological obstacle to the acceptance of fresh milk: high rates of lactose intolerance.[12]

Let us briefly consider the assertion that premodern Chinese societies eschewed dairy, which sounds suspiciously like the kind of totalizing claim that nutritional imperialists made about wheat and meat. One early piece of evidence that supports it comes from the *Yellow Emperor's Inner Canon* (*Huangdi neijing* 黃帝內經), compiled around the first century BC. Its authors suggested that people from different regions were constitutionally suited to the consumption of different foods. Milk, in this schema, was for people from the north, probably meaning the frontier region around the Ordos Loop of the Yellow River and what is today Mongolia.[13] From this perspective, dairy eating was a regionally bounded habit, appropriate for people whose constitutions were compatible with it, but not for the people of the other regions, including the center, considered the heart of Chinese civilization.

This distinction turns out to have been less absolute in practice than the *Inner Canon* made it seem. Han Chinese did, in fact, milk livestock and consume dairy before modern times. As early as the Former Han Dynasty (207 BC–9 AD) the imperial household employed an official who collected mare's milk for "mare's-milk wine."[14] The sixth-century AD agricultural text *Essential Techniques for the Welfare of the People* (*Qimin yaoshu* 齊民要術) includes careful instructions for milking cows and goats or sheep and for producing a fermented beverage from their milk.[15] As the historian Françoise Sabban and the biochemist and historian H. T. Huang have shown, Chinese literati demonstrated a sophisticated knowledge of milk and its properties and consumed a range of dairy foods that included kumiss made from mare's milk, goat's milk drunk raw, and sweets made from water buffalo milk.[16]

Some scholars have suggested that Chinese enthusiasm for dairy was short-lived and geographically bounded, characteristic mostly of periods such as the Tang dynasty (seventh to tenth centuries AD) and Yuan dynasty (thirteenth and fourteenth centuries AD), which saw extensive intermingling and intermarriage between Han Chinese and the peoples of Central Asia and the steppe—and even then, mostly among people living in the northern part of the empire. But more recent research provides intriguing evidence that even southerners living much later had a taste for dairy and knew how to handle it. The historian Miranda Brown has explored a sixteenth-century recipe book compiled by a southerner and found that it featured cheese as often as it did tofu, including a recipe for cheese-stuffed fish and even one for deep-fried breaded milk cakes that sound a bit like modern mozzarella sticks.[17] Far from considering dairy abhorrent or uncivilized, it seems, some premodern Chinese eaters enjoyed the occasional cheesy treat.

Chinese doctors also systematically investigated milk's therapeutic properties and deployed it as a medicine. Judging from its prominence in formularies, cow's milk featured frequently as an ingredient in drug prescriptions, and *materia medica* books detailed the uses of horse, camel, sheep or goat, human, pig, and dog milk as well. Few lactating mammals, it seems, escaped the healer's gaze. In a seminal eighth-century formulary that practitioners of Chinese medicine continue to consult today, cow's milk is among the top 25% most commonly mentioned ingredients.[18] It took many different medical forms. Its earliest uses were topical, as an element of a poultice or a rejuvenating face cream, for example. As time went on, healers came to use it more often in decoctions, boiling it with herbs and minerals to make a curative drink. In some formulas, they used milk to help bind drug ingredients into pill form or as a way to wash down a dose of pills.

When it came to the conditions it treated, milk was similarly versatile. It was indicated for ulcers; excessive sweating; pain in the chest, legs or lower back; mental agitation and depression; vomiting in children; and much more. The compilers of medical books saw milk as particularly appropriate for conditions involving excessive heat, as they considered most milks cooling by nature. They recommended milk-based pills to counteract the effects of heat-producing foods such as wheat noodles or wine, for example. As a singularly fortifying tonic, milk also appeared often in remedies for disorders associated with wasting, thirst, and depletion and was recommended for more fragile bodies such as those of the very young, older people, and convalescents.[19]

Chinese healers also prized milk for its effect on the gut. In some cases it served as a laxative that accelerated digestion; inducing diarrhea was the sign of successful treatment with some milk-based decoctions.[20] Incidentally, this seems to have been a common perception in many parts of the world, not just China. Older women in Malaysia and Indonesia recount using milk in this way, and in Ayurvedic medicine in South Asia milk was thought to facilitate digestion, "even being used as an enema or purgative." In southern Europe, too, a Renaissance scholar wrote of milk that it served to "loosen the bowels." In Chinese classics, milk was sometimes deployed to achieve something like the opposite effect: to restore regularity in crises involving uncontrolled defecation.[21]

Nor was milk used only therapeutically. Li Shizhen, the sixteenth-century author of one of China's most famous medical works, the *Compendium of Materia Medica* (*Ben cao gang mu* 本草綱目), emphasized milk's prophylactic qualities in a little poem he appended to the cow's-milk entry of his magnum opus. Li called milk the "liquor of the immortals" (*xianjia jiu* 仙家酒) and declared that if one could drink a *sheng* 升 [about one liter!] or more each morning one could "reverse aging and return to youth."[22] In the past, historians inform us, most Chinese people who drank milk preventively like this did so only in the winter because it spoiled so easily in hotter weather.[23] This observation hints at an important material limitation of premodern consumption and a significant change in the modern period: before pasteurization and widespread, portable refrigeration, *fresh* milk was drunk primarily in places geographically close to milk-producing animals.

In sum, the picture of milk in premodern China that has emerged from a closer engagement with culinary and medical sources diverges dramatically from the conventional depiction of China as a lactophobic culture. Who are these cheese-making gourmands quaffing milk to stay young? It turns out that not only were premodern Chinese familiar with milk, but they also had a nuanced understanding of how to process and use it in that they recognized that the milks of different animals had different physiological effects and were appropriate for some stages of life and some health conditions but not for others. Cow's milk could scour the gut, loosen stools, fortify the weak, or counteract a heating meal or drink. It could cool and moisten. Importantly, and in contrast to narratives that have developed since the concept of lactose intolerance was invented, Chinese people themselves seem *not* to have seen either certain people, or themselves as a collective, as being unable to handle dairy.

Milk was not foreign to premodern China. So thoroughgoing was its modern erasure from memories of the past, however, that scholars have only recently begun to rediscover its diversity and abundance in Chinese cuisine and medicine before the age of imperialism.

Early Twentieth-Century Lactomania and the Industrial-Milk Gap

How, then, did modern people come to forget about China's milk heritage—to imagine that dairy had always been alien to Chinese cuisine and repulsive to Chinese eaters? This change began in the early twentieth century, and like the example of meat, it grew from a kernel of empirical observation. Just as Europeans and Americans were eating a lot more meat than Chinese people *at that time*, they were also drinking a lot more fresh milk. This was a relatively new habit, as historians have shown. In the United States, most dairy had been consumed in the form of cheese or butter as late as the mid-nineteenth century. When Americans imbibed dairy in liquid form it had usually been buttermilk, a byproduct of butter-making.

But as the historian E. Melanie DuPuis has written, the changes that urbanization wrought in women's lives precipitated a decline in breastfeeding. In cities, women lived in households that were more isolated and independent than the ones typical of the countryside, and they had no equivalent to the network of female helpers who supported rural women in their work, childcare, and nursing. Demand thus grew for a less time- and labor-intensive form of nourishment for infants and young children, and a year-round supply of fresh cow's milk developed to meet that demand.[24] At the same time, advances in refrigeration and transportation technology, not to mention the advent of pasteurization, made fresh milk safer than it had been in the past. Large-scale dairy farms began to emerge, and consumers in burgeoning industrial cities increasingly had access to milk for drinking. A US Department of Agriculture report in 1909 estimated that the average American family of five consumed about a quart of fresh milk daily and observed that "milk and cream together furnish 16 percent of the total food of the average American family."[25]

A few illustrations serve to highlight the extraordinary faith that milk inspired in nineteenth- and twentieth-century scientists. In the 1870s and 1880s, some American and English doctors were promoting the intravenous injection of cow's milk "in the later stages of cholera, enteric fever, phthisis, and pernicious anemia, as a substitute for the transfusion of blood," the idea being that milk was highly nourishing and that milk injections were easier and safer than blood transfusions.[26] By the 1920s, nutrition scientists in the United States were exhorting children and adults alike to drink milk, calling it a complete food. Milk was complete from the scientists' point of view because it contained all the major amino acids, minerals, and vitamins that had been discovered in human bodies up to that point (cow's milk is not naturally rich in vitamin D, but vitamin D had yet to be discovered when the milk craze began). For the New Nutritionists, obsessed as they were with achieving optimal levels of micronutrients, that made milk perfect.

So perfect did it seem, in fact, that it formed the basis of a new unit for measuring nutritional value, the NEM (Nahrungs-Einheit-Milch in the original German, which English speakers sometimes translated "nutritional equivalent of milk"). The NEM never succeeded in replacing the calorie as its Austrian creator hoped it would, but it did underpin food-relief efforts after the First World War, when several hundred thousand Austrian children ate rations measured in NEMs.[27] It would be difficult to find a more perfect emblem of nutritional imperialism than this: an attempt to turn a foodstuff important to the diets of well-to-do Westerners into the standard for measuring the nutritional value of *all* food. Milk, it seems, had become virtually synonymous with nourishment.

Meanwhile, although as we have seen, Chinese people were knowledgeable about dairy, the scale of production and consumption in early twentieth-century China was nowhere near as extensive as it was in industrialized Western countries. The scope and nature of dairy production in China's imperial period have not been adequately researched and remain unclear. It is apparent, however, that by the early twentieth century, apart from relatively well-to-do urbanites, people in China were not drinking as much milk as people in Western Europe and North America.[28]

Thanks to dietary determinism and imperialist hubris, this relatively new divergence came to be seen, like meat eating and wheat eating before it, as a distinguishing characteristic of two civilizations from ancient times. Just as Europeans and Americans had divided the world into two tiers of wheat eaters and rice eaters, meat-eating and vegetarian nations, they perceived a similar divide when it came to milk. In the 1920s American public-health officials rhapsodized, "The races which have subsisted on liberal milk diets are the ones who have made history and who have contributed the most to the advancement of civilization."[29] The US National Dairy Council concurred: "The people who have achieved, who have become large, strong, vigorous people, who have reduced their infant mortality, who have the best trades in the world, who have an appreciation for art, literature and music, who are progressive in science and every activity of the human intellect are the people who have used liberal amounts of milk and its products."[30] Such triumphant rhetoric might not be surprising from milk-boosting Americans, but it was hardly confined to them. Gan Rudeng, the author of a 1924 article in the Chinese periodical *Farming Monthly* (*Nongshi yuekan* 農事月刊), similarly asserted, "From history and empirical observation, science has proven that all of the races that drink ample cow's milk or eat ample eggs or meat are more developed. . . . In the nations of North America, the milk industry is most widespread, and the amount of milk produced is very great. And today's world leaders in physical, intellectual, and moral education all love the nurturing [effect] of cow's milk."[31] Because Gan did not perceive his country as a *properly* milk-drinking nation, he saw "stimulating the dairy industry" there as "truly one way of saving China."[32] Over ten years later, a 1937 article in the same magazine made almost exactly the same claim: "Today the countries with the loftiest culture are precisely those that consume the most cow's milk and milk products. Inferring the reason, it is simply that the nutrients of cow's milk are abundant and easily suit [their] citizens' nourishment. Their bodies are strong and brains healthy, and they are bold and enterprising, and therefore their scholarship is advanced, their institutions are flourishing, and their culture is superior because of it."[33] As these examples show, some Chinese observers were as convinced as any European or American writer that a milk gap had shaped a power gap.

Older Chinese milk ideas and practices were not forgotten right away, if early advertisements in Chinese newspapers are any indication. The newspaper *Shun Pao* 申報 (called *Shanghai News* in English) featured ads starting around 1912 for nutritional supplements that combined cow's milk with malt extract, which, they declared, could especially help older people whose *qi* was depleted and whose blood was weak. A cow's-milk ad from 1917 said the product would moisten the skin, relieve agitated thirst (*fan ke* 煩渴), and replenish a person from the depletion caused by labor—all functions that the sixteenth-century *Compendium of Materia Medica* had credited milk with.[34] A 1915 article called "Ways of Nourishing Life through Cow's Milk" (*Niuru yangsheng zhi fa* 牛乳養生之法) published in *Progress* (*Jinbu* 進步), a general-interest magazine produced by the Chinese YMCA, combined old and new practices in a startling way. It gave a nod to the traditional use of milk in healing by stating that readers already knew that dairy was the most important thing for sick people to eat. But, it went on, they might not know that "people can also use cow's milk to nourish life even if they are *not* ill. Its benefits are greater than fish or vegetables and fruits." The author then offered nine methods for incorporating milk into one's diet that were supposedly "convenient to put into practice," "extremely compatible with the way of cultivating life [*yangsheng fa* 養生法]," and derived from scientific experiments.

Mentioning *yangsheng* tied the recommendations to the classical dietetic tradition; mentioning scientific experiments invoked biochemists' authority. While the author's insistence that "cleansing your intestine with milk every weekend will help you nourish life" fit nicely with dairy's premodern role as a laxative, the amounts he recommended drinking were truly on a modern scale: Some of the methods involved downing more than four liters (a gallon) of fresh milk in a day.[35]

By the time nutrition scientists added their voices to the conversation in the 1920s, however, milk had come to be seen as another thing China lacked. True, cities such as Shanghai had a thriving, growing dairy sector supplying what had become a fashionable habit among the urban elite. Lu Xun's diary, for example, shows that the famous writer occasionally went out to drink cow's milk with friends the same way a group today might go out for coffee or beers.[36] But the relative size of the dairy industry when compared to countries in Europe and North America was miniscule, and regular dairy consumption was impossible for those of modest means. That was what the scientists worried about. As the Yenching University biochemist William Adolph concluded in the Chinese-language journal *Science* (*Kexue* 科學): "'Every day one cup [*zhong* 鍾] of cow's milk' is the chant of health experts in every country. If we want the 400 million people of China all to take cow's milk as a food, such a circumstance is definitely impossible. But since [milk's] connection with the health and physique of the nation is so tight, we must loudly and urgently call out 'Develop cow's milk!' to urge one another on."[37] In the meanwhile, as they waited for China's dairy industry to mature, the scientists saw promise in soy milk as an alternative, as Jia-Chen Fu has chronicled.[38] Chinese dairy peddlers, sensing the zeitgeist, stopped plugging their product as a familiar remedy—a *qi* and blood booster that relieved thirst and moistened the body—and started promoting it instead as a way to increase China's fitness in the Darwinian struggle for national survival.[39]

Just as modern rhetoric concerned with wheat and rice conveniently ignored the large proportion of Chinese people who ate wheat, the new rhetoric concerned with milk ignored dairying traditions outside of Europe and North America among people whom Western writers did not consider large, strong, or vigorous. This meant overlooking not only China's "savoir-faire oublié," as Françoise Sabban has called it, but also India's even more prominent heritage of milking.[40] It also projected the milk-saturated condition of early 1900s United States, where there was purported to be one dairy cow for every four people, backward to the dawn of Western civilization.[41] Recall that Hsien Wu had declared milk drinking a European trait from at least the time of the Christian Bible. He was following, in this account, erroneous claims by dairy boosters that drinking fresh milk was an ancient staple of Judeo-Christian civilization, prominent since biblical times.[42]

The fact that in the early twentieth century Chinese people did not produce or consume milk in the ways that resembled their European and American contemporaries does not mean that they were ignorant of milk or averse to it, or that they didn't use milk themselves in diverse ways. Industrial-scale production and consumption of fresh milk is not the only way to incorporate dairy into society, but that was increasingly how scientists and policymakers thought about it.

Cultivating Modern Industrial Milk in Chinese Communities

Throughout the twentieth century Chinese nutrition scientists, officials, and entrepreneurs duly worked to cultivate the new kind of milk production and consumption that Euro American states had pioneered. Although the vagaries of twentieth-century history kept the path from being smooth or linear, by the twenty-first century mass-produced fresh milk was firmly embedded in Chinese economies and diets. This section explores the contours of that transformation.

The historians Wang Fengzhan and Yu Xinzhong have traced China's modern dairy industry back to the foreigners who settled in treaty ports such as Shanghai and Tianjin starting in the mid-nineteenth century. Some of them brought cows from abroad and kept the animals in the yards of their residences to supply their own milk needs. They sold any surplus to their Chinese neighbors. Enterprising Chinese farmers living near the treaty ports began to breed their own work cattle with the newly imported, highly productive milk cows. In the winters—the traditional season in which Chinese consumed milk as a dietary supplement—the farmers would sell the milk, and in the spring and summer they would use their hybrid cows for planting and harvesting labor. Typically such a farmer might have anywhere from one or two to a dozen cows.[43]

Many such small-scale operations did not survive long into the twentieth century. In 1897 the government of Shanghai's international concession adopted a licensing system for anyone wanting to sell milk to concession residents. The requirements for obtaining a license—minimum standards for the size, condition, and maintenance of pens; maintaining sanitary milking and storage equipment; reporting illnesses among animals or people at the dairy—aimed to protect public health, but achieving them was too costly for most small producers. By the first decade of the 1900s, Beijing likewise began regulating dairies, and years later, when the Guomindang secured power, they announced a similar set of rules that would shape the dairy industry nationwide starting in 1928. Larger-scale, more profitable dairies that were able to meet the new standards flourished; small-scale family ventures struggled. Although the dairy sector grew impressively over the course of the Republican period, by the time it ended in 1949, regular fresh-milk consumption was still a "habit for a happy few," as Françoise Sabban has written.[44]

From the perspective of international public-health experts, China's (comparative) dearth of dairy was a problem. In 1935 the League of Nations Health Committee delegated a technical commission that met in London to come up with what it presented as a universal, global set of recommendations for daily diet. The scientists serving on this commission hailed from Austria, the United Kingdom, France, Italy, the Scandinavian states, the Soviet Union, and the United States. None came from China, in other words, or other parts of the world where people of European descent did not constitute a majority. The chairman of the body that published the committee's comprehensive report in 1937 was the Viscount Astor of Great Britain, who happened to be the "Chairman of the Milk-in-Schools Advisory Committee (under the Milk Marketing Board)" in his home country. The League of Nations' recommendation about milk was what one might have expected given the participants' backgrounds. In all countries, they asserted, milk "should represent a large proportion of the diet at every age."[45] How large? Pregnant and nursing women were supposed to get 660 (27.5%) of their 2,400 daily calories from milk. For children the percentage was higher, with about half of a child's total daily calories up to the age of 14 ideally coming from milk.

A year later the League of Nations published a report called *Nutrition in Various Countries*. It reviewed the global state of nutrition but its purview was limited; the only Asian country examined in the report was Siam (Thailand).[46] Understandably, Chinese nutrition scientists felt that it would be a good idea to assess whether the supposedly universal guidelines that the league was propounding were appropriate for China. The China Medical Society (*Zhonghua yixuehui* 中華醫學會) accordingly convened a Special Working Group on Nutrition consisting of Hsien Wu, his colleague Hsiang-ch'uan Hou from Shanghai's Lester Institute, and Siu-feng Huang and William Adolph, both scientists based at the Peking Union Medical College. They observed, "Cow's milk is not very common in China" and noted that "animal-derived foods" were quite expensive there; consequently, they were not as insistent as the League of Nations committee about the necessity of milk. They did present it as an ideal for people who had sufficient means, however. They recommended that if a family could afford it, about a quarter of the total calories in an 8-to-10-year-old child's diet should come from cow's milk. According to their recommendations, this would have been about 700 hundred milliliters of milk daily, about one half the proportion that the League of Nations committee had recommended but still a significant amount.[47]

Almost 10 years after the League of Nations issued the original global report, its successor, the United Nations, produced a new one, the Food and Agriculture Organization's first *World Food Survey* in 1946. The report, hastily written in the immediate wake of the Second World War, estimated prewar nutritional status around the world. Its authors asserted that a relative few regions and countries had enjoyed balanced diets with adequate calories, namely "North America, Oceania, Argentina, the British Isles, Scandinavia, the Netherlands, Switzerland, and Germany."[48] Other places should aim to meet those same standards, they thought, but doing so would be a challenge—particularly when it came to minimum milk requirements in East Asia. They estimated that 10%–13% of the calories in an adult diet should come from milk and lamented, "to supply the additional quantities of milk and milk products required in the Far East to attain the desirable minimum . . . will be outside the bounds of practicable possibility" for a long time, "since the prewar consumption of milk in that part of the world was almost negligible." The accompanying table accentuated this point for China specifically. It estimated how much China would need to increase the supply of different food categories in order to meet the UN's global nutrition targets by 1960. For other categories the needed increase ranged from a modest 15% for cereals and sugars to an ambitious 327% for fruits and vegetables. But the required increase in "milk, fluid equivalent"—5,650%—eclipsed the other gaps and seemed truly unreachable.[49]

As it turned out, however, China was able to go a long way toward catching up by the turn of the twenty-first century, a trajectory I will briefly sketch for China's two different manifestations since 1949—namely the People's Republic and Taiwan. Milk availability and consumption developed in a distinctive way in each of these settings, which is not surprising given their different postwar trajectories. Here, I will consider their paths separately.

In 1950s Taiwan, the same dynamics that had brought in thousands of tons of American wheat, as described in chapter 2, brought a major influx of American dairy as well. American producers had been churning out dairy products in amounts that domestic markets could not absorb.[50] To rescue farmers from the economic losses that accompanied this oversupply, the US government purchased large amounts of milk and promoted its consumption by formerly non-dairy-consuming populations, both domestically and abroad. In 1956 Congress authorized $75 million to be spent over 3 years "to increase the consumption of fluid milk by children," a program that continued to be renewed through the 1960s.[51] Milk and butter became central items on American school-lunch menus. Internationally, Public Law 480 made the United States the world's largest supplier of food aid starting in 1954. The law was designed primarily to provide an outlet for excess agricultural production; its designers also hoped that providing humanitarian assistance in the form of heavily subsidized American food would promote goodwill toward the United States and strengthen the country's allies, including Taiwan.[52]

Thus in 1956 Taiwan agreed to buy 5,000 metric tons of discounted milk products from the United States for the equivalent of $1.5 million. This was the beginning of a long period during which cheap American dairy products would enter the country and foster a growing demand for milk while simultaneously suppressing, in some ways, the growth of a domestic dairy industry.[53] This American import altered eating habits and cuisine on the island. In the 1950s milk became for the first time the drink of Taiwan's schoolchildren, as happened in Japan around the same time for the same reason.[54] Among all Taiwanese, average milk consumption in 1965 was three times as much as it had been seven years earlier.[55]

Meanwhile, the politically isolated People's Republic of China (PRC) received no comparable influx. Dairy remained relatively rare there, as the historian Thomas David DuBois has detailed. Editorials in the state-run media decried the American dumping of milk powder on the rest of the world as a capitalist plot while simultaneously trumpeting milk production levels in the Soviet Union and Eastern Europe as evidence of the benevolence and efficacy of socialist governance.[56] The PRC came nowhere close to matching the production levels of its fellow Communist states, however; in 1958 China's average per-capita consumption of milk was "barely 1% that of the Soviet Union."[57] The subsequent great famine that lasted into the early 1960s and then the chaos and neglect of the Cultural Revolution reduced supply even further. Milk was reserved for urbanites, soldiers, and party cadres and was rationed even in the cities. It was not until the early 1970s, as the Cultural Revolution's early years receded and the PRC began to open to more international connections, that production began to stabilize and grow. One sign of this was the emergence of the journal *China Dairy Industry* (*Rupin gongye* 乳品工业), which began publication in 1973. It featured reports on the dairy industries of other countries as well as on efforts within China to increase production and consumption. The major changes, however, would have to wait until the death of Mao Zedong in 1976 and the rise of the reformer Deng Xiaoping as his successor.

In 1979, at the dawn of the reform-and-opening period that Deng Xiaoping's policies ushered in, *China Dairy Industry* could proudly report that China's farmers had produced "more than 46,000 tons" of dairy foods in the previous year, "the highest level since the founding of the PRC."[58] This may have been a lot by the standards of the early PRC, but it was still a very small industry compared to dairy giants like the United States; in 1980 American farmers produced more than 64 million tons of milk.[59] Still, it signaled that increasing the availability of milk products was among the goals of the new regime, an impression confirmed by the advent of new industry periodicals such as *China Dairy Cattle* (*Zhongguo nai niu* 中国奶牛) magazine, which started in 1983, and by state propaganda. Fresh milk featured in illustrations of the materially prosperous life that Deng's reforms promised. One poster trumpeting the way agricultural modernization would improve the lives of future generations, for example, centered on a young girl gazing happily at a bottle of milk (Figure 4.1).

Market liberalization did indeed cause a fast rise in milk production in the 1980s, and since 2000 the Chinese milk supply has grown even more sharply as ultra-heat-treated (UHT) shelf-stable milk has become widely available. China's entrance to the World Trade Organization in 2000 increased competition and prompted consolidation in the industry, with the result that a handful of Chinese companies began to dominate the domestic

Figure 4.1 PRC poster from the 1980s: "Nongye xiandaihua zan: weile zisun shenghuo meihao 农业现代化赞:为了子孙生活美好" [In praise of agricultural modernization: To better the lives of our children and grandchildren]. A little girl gazes at a bottle of milk (and notice the wheat spikes in the foreground). BG E15/685 (chineseposters.net, Landsberger collection). Used with permission.

market. By 2008 Chinese milk production was five times larger than it had been only eight years before.[60] China has not only adopted but has also taken the lead in industrial-scale milk production; it now has the largest dairy farms in the world, with tens of thousands of cows.

Demand has grown apace. Chinese consumers have rapidly increased their intake of fresh milk, drinking more than 3 times as much in 2010 as they had in 2000—and nearly 20 times as much as in 1970. Burgeoning Chinese demand has changed the global market for cow's milk.[61] Even a contamination scandal that dominated national and international headlines in 2008 did not appreciably dim Chinese enthusiasm for milk.[62]

It is clear, in this era of milk mania, that the early twentieth-century belief in milk as an emblem of modernity and fuel for national strength has lingered. A 1999 article in the *Shanxi Food Products Industry* (*Shanxi shipin gongye* 山西食品工业) magazine rehearsed the familiar themes about milk as a nutritionally complete food and ended with a reflection on milk's significance in nation building. The author praised Winston Churchill, the Thai monarchy, and the Japanese government for actively promoting milk consumption in their respective countries, particularly by children. Focusing on a correlation between milk drinking and average height, they wrote, "As everyone knows, after the war Japan brought up [the concept of] 'a glass of cow's milk to strengthen a nation,' and for a long time they had a large-scale 'children's milk plan.' This undoubtedly has had a decisive function in Japan's shaking off the 'short race' image they used to have."[63] The conviction that milk makes people taller is widespread in East Asian communities, where short stature is a stubborn concern, though as the anthropologist Andrea Wiley dryly notes, "The scientific literature does not provide strong support for this view."[64] In any case, the conclusion of the *Shanxi Food Products Industry* piece could easily have been written 60 years earlier: "The consumption of milk and the development of milk products are a basic policy that strengthens the nation's physique and advances national flourishing and development—we should give it a high degree of attention."[65]

Industry boosters have not been the only ones advocating for Chinese citizens to drink more milk. The government has likewise strongly promoted this dietary transition. The most visible symbol of this high-level advocacy was then-premier Wen Jiabao's visit in 2006 to an industrial dairy farm in Chongqing, where he famously declared, "I have a dream that every Chinese person—and children foremost—will be able to drink a *jin* 斤

[about 500 milliliters, or a pint] of milk every day."[66] As befits a "dream," Wen's target was audacious. Even in the United States, where per-capita milk consumption has been many times higher than in China, people have been drinking only about 200 milliliters a day on average.

Official Chinese dietary recommendations, presented as targets rather than dreams, have been only slightly less ambitious. In 1997, under the auspices of the Ministry of Health, the Chinese Nutrition Society and the Nutrition and Food Safety Research Institute of the Chinese Preventive Medicine Academy created an expert committee to revise the first set of dietary guidelines that the Nutrition Society had produced in 1989. The first guidelines had featured eight general principles that combined classical dietetic with common-sense recommendations: "Eat a variety of foods; maintain an appropriate state of hunger and fullness; use an appropriate amount of oils and fats; balance rough and refined foods; limit the amount of salt eaten; eat only a little of sweet things; drink alcohol in moderation; eat three equal meals."[67] The second set, however, added an emphasis on consuming dairy, specifying that a Chinese resident should consume 300 grams of milk or dairy products daily. The committee accentuated the point with a "diet pagoda" graphic communicating more succinctly and directly what an ideal diet was supposed to look like. In it a carton and glass of milk feature prominently (Figure 4.2).

This emphasis on dairy has persisted in subsequent revisions of the guidelines and has even grown more pronounced. A 2016 revision introduced a "balanced diet plate" graphic in which the four food categories (grains, vegetables, fruits, and fish-meat-eggs-and-beans) assumed the shape of a yin-yang symbol (Figure 4.3). The diagram elevated dairy by giving it an independent place. Above and apart from the plate sits a cup full to the brim of fresh milk, emphasizing not only that dairy has a uniquely important place in a Chinese diet, but also that drinking fresh milk—not eating cheese, yogurt, or ice cream—is the form that dairy consumption is supposed to take.

These official recommendations and their accompanying illustrations demonstrate how standards originally crafted in the United States have continued to influence Chinese dietary guidelines. Although the original diagram that accompanied the 1997 dietary guidelines featured a traditional Chinese style of architecture—the pagoda—both its content and layout bear an unmistakable resemblance to the US Department of Agriculture's food pyramid from the early 1990s (Figure 4.4).

Figure 4.2 Chinese food guide pagoda (*Zhongguo jumin pingheng shanshi baota* 中国居民平衡膳食宝塔), 1997 version. Chinese Nutrition Society.

Figure 4.3 Balanced diet plate for the Chinese (*Zhongguo jumin pingheng shanshi canpan* 中国居民平衡膳食餐盘), 2022 version. Chinese Nutrition Society.

Apart from the curved roofs on the pagoda, there is little to distinguish the two graphics from one another. Even the location of dairy products on the Chinese graphic is the same: second tier, left-hand side. Although the justification for including dairy has changed—it is said to reduce the risk of chronic disease rather than to boost strength and initiative—the faithfulness of the imitation suggests that American dietary ideals have a normative influence similar to what they exercised in 1920s China.

As in the case of Taiwan importing discounted American milk products, the motivations fueling China's contemporary milk boom have been as much economic as nutritional. The PRC government has seen increasing dairy production as a way to achieve economic, social, and environmental goals. In 2002, faced with catastrophic dust storms that originated in the parched terrain of Inner Mongolia and then ravaged Beijing, the government coerced and compelled Inner Mongolia pastoralists to become dairy farmers instead. Although they would still be caring for animals, they would be tending dairy cows in concentrated animal feeding operations (CAFOs), which requires less land than traditional multispecies pastoralism. On the surplus land where their horses, sheep, goats, and cows had formerly grazed, the farmers were to plant trees and shrubs to limit erosion and

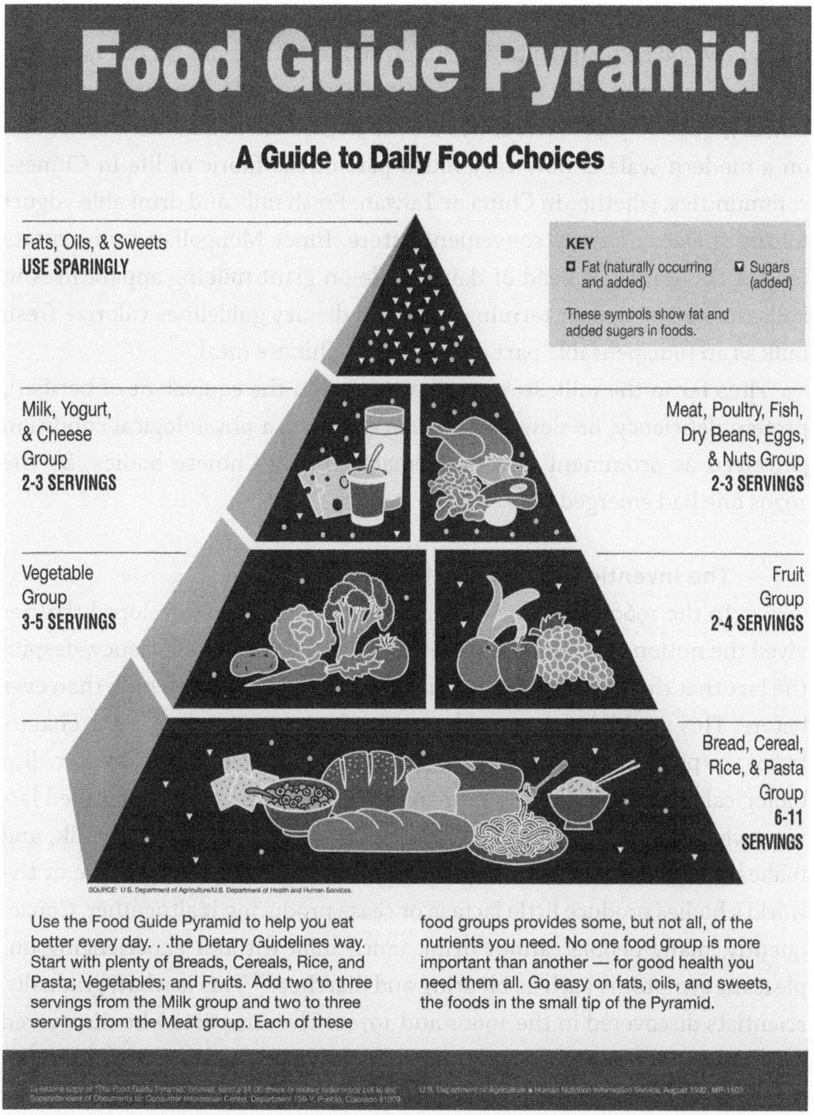

Figure 4.4 The Food Guide Pyramid of the United States Department of Agriculture, 1992. Dairy products are located on the left-hand side of the second tier from the top, just as in the later Chinese food pagoda. Internet Archive URL: https://web.archive.org/web/20170918213032/https://nutritionhistory.nal.usda.gov/download/1759165/PDF.

tamp down dust storms. Not coincidentally, minting so many new dairy farmers would help ramp up milk supply to meet ambitious official targets.[68]

China may not have had much of a dairy sector in the early twentieth century, but 100 years later it was a behemoth. Milk in a modern sense and on a modern scale is now very much part of the fabric of life in Chinese communities, whether in China or Taiwan. Fresh milk and drinkable yogurt fill the shelves of every convenience store. Inner Mongolian farms rotate tens of thousands of head of dairy cattle on giant milking apparatuses in halls the size of airport terminals. Official dietary guidelines valorize fresh milk as an indispensable part of a balanced Chinese meal.

Thus far in the milk story we have not seen the equivalent of beriberi, protein deficiency, or slow metabolism—that is, a physiological condition perceived as prominent and problematic among Chinese bodies. By the 1970s one had emerged.

The Invention of Lactose Intolerance

In the 1960s and 1970s a pernicious new concept developed that revived the notion that Chinese people had a dairy-related deficiency, despite the fact that they were producing and consuming more fresh milk than ever before. This was the concept of *lactose intolerance*. It is tempting to characterize the phenomenon of lactose intolerance as a discovery. It is, after all, a biological reality. All healthy human babies produce an enzyme called lactase, which breaks down lactose, the main sugar in any kind of milk, and makes it digestible. But after early childhood, two-thirds or more of the world's bodies produce little lactase or cease producing it altogether. Consequently, many people cannot drink much milk without experiencing unpleasant symptoms such as bloating and diarrhea.[69] This is what, gradually, scientists discovered in the 1960s and 1970s. The scientists who discovered this interesting variation in human biology, however, were mostly of European descent and accustomed to thinking of Western European bodies as normative, and they framed it as a medical condition. Pathological designations accreted around the response of healthy adult bodies to fresh milk: *lactose intolerance, lactase deficiency, lactase insufficiency, hypolactasia*. That lactase production usually diminishes after infancy was a discovery, in other words, but "lactose intolerance" as a medical problem was an invention.

The discovery that some bodies digested dairy more easily than others was a direct consequence of the postwar programs to offload excess

European and American dairy in places where people had not previously drunk large amounts of milk on a regular basis. Thanks to the new domestic and international government initiatives, African Americans, Asian Americans, and many people in East and Southeast Asia began to imbibe fresh milk, and a significant proportion of them began simultaneously to experience symptoms of discomfort. Observing this prompted scientists to study the phenomenon. They wanted to understand what was wrong—as they framed it—with the people whom milk made sick.

The first studies to highlight racial difference in lactase levels appeared in the United States in 1965 and 1966. One of these was the report of Theodore Bayless and Norton Rosensweig, from Johns Hopkins University, in the *Journal of the American Medical Association*. They studied 20 White and 20 Black inmates at a prison, all in good health. They tested for lactase deficiency by having the men ingest 50 grams of lactose and then provide blood samples six times over the subsequent two hours, testing for a rise in glucose level that would indicate the presence of lactase. Bayless and Rosensweig also biopsied and assayed the men's jejuna, a part of the small intestine, to assess lactase levels directly. They concluded that most of the Black inmates were lactase deficient.

Interestingly, the overwhelming majority of Black inmates had themselves reported, before any blood was drawn or organs biopsied, that milk altered their digestion. The men did not phrase it as "milk intolerance," a medical condition. Rather, they saw the loose stools associated with milk consumption as a characteristic of the milk itself, viewing it as "a mild laxative." As the study's authors reported, "Many extolled [milk's] virtue as a producer of 'regularity.'"[70] In this, their perspective resembled those expressed in classical Chinese medicine and Ayurveda, where milk often acted as a useful purgative.

For the White researchers, however, the phenomenon implicated low lactase levels, which they saw as "inborn errors of metabolism." They speculated that this defect might be inherited.[71] Bayless and Rosensweig's study sparked more research into racial differences in lactose digestion and garnered attention in the popular press.[72] At first scientists sought to explain the evolutionary origins of lactose *in*tolerance. Like many researchers after them, Bayless and Rosensweig assumed that intolerance was an aberration from the dominant state of humankind and speculated that it might be a "recessively inherited condition" that required "homozygosity for appearance of the trait." They proposed an explanation that is the inverse of what scientists now believe, hypothesizing that over many generations, some people may have *become* lactose intolerant because they did not drink milk after weaning. In a milk-free environment, they reasoned, producing lactase would have wasted energy and resources, so when random mutations yielded bodies that stopped producing lactase, those bodies would have had a slight competitive edge.[73]

Scientists struggled to grasp that a trait unlike the one found in people of northern European descent could, in fact, be more typical. As research began to accumulate in the late 1960s indicating that not only Black Americans but also Asian immigrants, Ugandans, and Thais and other Southeast Asians produced little or no lactase as adults, a fierce debate developed over whether this "abnormality" was genetic or acquired.[74] Some reasoned along the same lines as Bayless and Rosensweig that lactase-deficient populations had mutated away from the original template over many generations, while others suggested that whether an individual continued to produce lactase after weaning might, in fact, simply reflect whether they continued to consume milk.[75] It was the geographer Frederick J. Simoons who, in 1970, first proposed that lactase non-persistence might have been the original state and lactase persistence the mutation. By looking at what was then known about the distribution of lactose intolerance in milk-producing and non-milk-producing societies, he concluded that lactase persistence had probably developed as a beneficial mutation in milking societies. As he observed, "It seems more promising to view the Western pattern, of high levels of intestinal lactase and of lactose tolerance throughout life, as the aberrant one which must be accounted for. This is because the Oriental pattern is like that of almost all land mammals . . . and the Western pattern is strikingly different."[76]

This viewpoint came gradually to be the scientific consensus. By 1981 the director of the National Institute of Child Health and Human Development at the US National Institutes of Health could write definitively in a volume on lactose digestion that "lactose intolerance is a normal physiological condition, shared by every adult animal except for certain ethnic and racial groups in man, most of whom are located in the United States or in northern parts of Western Europe. It is lactose tolerance rather than lactose intolerance, as initially assumed, that is most significant. The ability to digest lactose is the abnormal state, while lactose intolerance is the usual, normal condition in the human adult world population."[77]

Consensus did not, however, change the way scientists and physicians framed lactase non-persistence; they continued to write about it as a deficiency. Although the introduction quoted above acknowledged that "lactose tolerance . . . is most significant," most of the book it prefaced was devoted to chapters treating lactose *intolerance* as a problem, including chapters on diagnosis, screening, and clinical consequences in adults and children, and promoting the development of low-lactose dairy products so that those suffering from this deficit could still consume milk.

At the same time that clinicians, geneticists, geographers, anthropologists, and others were puzzling over the evolutionary origins of lactase non-persistence, some grew concerned that lactase non-persistence threatened the effectiveness of food aid provided to poor populations, both domestically and internationally.[78] In their 1966 study, Bayless and Rosensweig had mentioned, almost as an aside, that "milk intolerance may also have some sociologic significance since some welfare programs provide powdered milk for impoverished recipients, many of whom are Negro and who might be expected, on the basis of this study, to have lactase deficiency."[79] By the early 1970s this concern had developed into a debate. Some argued that giving milk to a malnourished lactose-intolerant person would not improve their state of nourishment because they would simply refuse to drink it, or worse, that it would exacerbate the problem by giving them diarrhea.[80] In malnourished children in poor countries, the extra loss of fluid could even be fatal.[81]

In retrospect, this moment might have opened an opportunity to reevaluate the ideas about a universally healthy diet that had emerged in the early twentieth century and had been built on Euro-American cuisine and observations of White bodies. Could other sorts of diets, perhaps even diets without dairy, promote health and growth and vigor? In the end, however, milk was too entrenched—in the global economy and in the convictions of nutritionists and policymakers—for observations about metabolic variability to unseat it. The United Nations' Protein Advisory Group wrote in 1972, "Milk is considered a virtually complete food and in the developing areas of the world, where protein deficiency is widespread and protein-calorie malnutrition [is] a serious childhood problem, the use of milk for child feeding programs is strongly advocated by all nutrition experts."[82] With backing from such international authorities, those arguing to retain milk as part of domestic and international assistance programs won the day. Milk powder has remained an important part of international food aid—more prominent when donor countries have agricultural surpluses to shed, less prominent when those surpluses dwindle—to the present.[83]

In China, as the idea that milk is a universally good food gained broader acceptance, the idea that lactase non-persistence is a pathology accompanied it. The Chinese terms coined for the phenomenon are pejorative and medical, just like the English originals: *rutang bunai zheng* 乳糖不耐症, "intolerant-of-lactose syndrome"; *rutangmei quefa* 乳糖酶缺乏, "lactase deficiency." They express, like the English originals, the conviction that lactose-tolerant bodies are normal and that the far more numerous others are deviations from that norm.

Since the 1980s, as milk has become more prevalent in China, lactose intolerance has increasingly been presented as a problem not just for dairy producers but for public and individual health as well. Dairy industry representatives have allied themselves with experts in nutrition and preventive medicine, and articles on lactose intolerance have begun to appear with greater frequency in medical literature. Increasingly, state resources have been devoted to developing low-lactose dairy products to help Chinese consumers overcome this perceived problem. In 2001, the Chinese Nutrition Society and the China Dairy Industry Association jointly organized a conference on lactase and low-lactose dairy products with DSM, a Dutch manufacturer of lactase. The conference brought together researchers from respected institutes such as the Chinese Academy of Preventive Medicine and from provincial medical colleges. Their participation constituted a tacit endorsement of the idea—congenial to the dairy industry and DSM—that lactose intolerance is a problem requiring attention and that adding lactase to dairy products (not, for example, consuming less milk) is its best solution.[84] A pair of authors from DSM's Chinese affiliate proclaimed in *China Dairy Industry* (*Zhongguo ruye* 中国乳业) magazine in 2012 that China would be unable to reach Premier Wen's milk-drinking dream until the country overcame the problem of lactose intolerance.

More recently, the belief that Chinese lactose intolerance impedes public health has reached what one might see as its logical culmination: On April 24, 2012, scientists at Inner Mongolia Agricultural University introduced to the world the first cow genetically engineered to produce low-lactose milk. Her name in Hanyu Pinyin was *Lakesi* 拉克斯, variously rendered in the English-language press as "Lucks" or "Lakes," and the scientists who created her predicted that she would begin producing China-friendly milk within a few years, after birthing her first calf. "Her birth," a celebratory account proclaimed, "brings glad tidings to the lactose-intolerant consumers of the world."[85] It is unclear what happened to Lucks; I have seen no follow-up coverage in the Chinese press, and no one has answered my inquiries to Inner Mongolia Agricultural University. But research published more recently suggests that the effort to genetically engineer a low-lactose cow continues.[86]

If one steps back to consider the context in which scientists discovered this difference in lactose metabolism, it is not hard to see the dynamics of nutritional imperialism at work. In the postwar period Americans and Europeans attempted to uplift poor nations of the world by providing them with copious amounts of milk, an effort that not coincidentally stood to economically benefit American and European farmers looking to sell their produce. The observation that many people in the recipient nations felt physically uncomfortable after drinking milk only confirmed the belief that there was something wrong with those people. It was up to Europeans and Americans to rescue them from poor health (not to mention small stature) not only by providing milk but additionally by devising solutions to overcome their physiological aversion to it. One might call it the "White Milk Burden," so similar were the prevailing assumptions to the White Man's Burden that underpinned nineteenth-century colonial conquests.

Research since the late 2000s has called into question the racialized understanding of lactose metabolism that emerged in the 1960s and 1970s, in which lactase non-persistence was presumed to be a condition that made "dark skinned people," as an op-ed in the *Washington Post* put it, categorically unable to drink milk.[87] As scientists have learned since then, lactose digestion is not a binary in which a person is either lactose tolerant or lactose intolerant; the human family displays, instead, a spectrum of lactase production after infancy.[88] In addition, differences in lactose digestion do not map neatly onto the racial and national lines that humans have drawn. Popular representations misrepresent this reality. A map accompanying a 2021 article in a nutrition-science journal (Figure 4.5), for example, implies that virtually all people in China are lactose intolerant while virtually no one in Australia, the United States, or northern Europe is.

The more nuanced truth is that while studies often report dramatic figures showing that 80% percent or more of subjects identified as Chinese tested positive for lactose intolerance, many Chinese people have no trouble digesting fresh milk—20% of 1 billion people is 200 hundred million

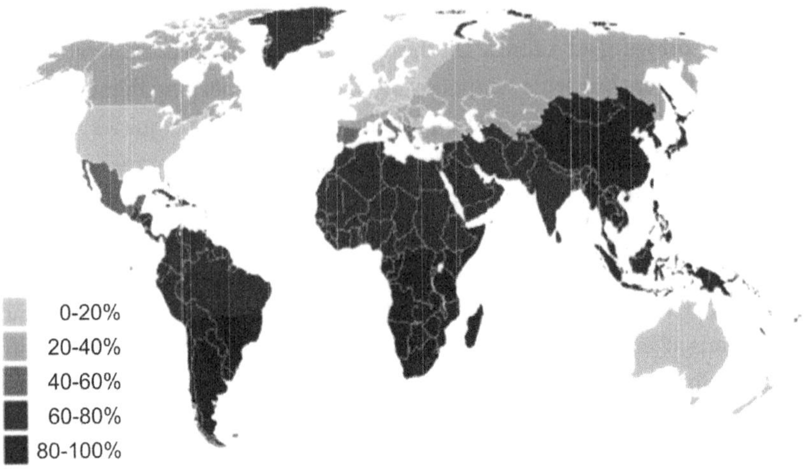

Figure 4.5 Illustration of the global prevalence of lactose intolerance, mapped onto national boundaries. This map misleadingly implies that lactose tolerance or intolerance is a homogeneous national trait. Moreover, if one removed the legend and left only the map, a viewer might assume the dark-colored areas indicate areas of the world dominated by White-majority imperial powers in the 1800s. Tatiana Capcanari et al., "Development of Lactose-Free Yogurt Technology for Personalized Nutrition," *Food and Nutrition Sciences* 12, no. 11 (2021): 1119, /https://doi.org/10.4236/fns.2021.1211082. Licensed under the Creative Commons Attribution International License (CC BY 4.0).

people, after all. Conversely, a significant minority of European-descended people cannot easily digest lactose.

Lactose digestion is, moreover, situational. Sometimes people who regularly drink a glass of milk without a problem fail a laboratory lactose challenge in which they ingest on an empty stomach an amount of lactose equivalent to a quart of milk. When they subsequently experience discomfort or meet the criteria for lactose intolerance on a hydrogen breath test, such experimental subjects are assigned to the "lactose intolerant" side of the supposed binary. Similarly, a person deemed "lactose tolerant" in the lab because a biopsy has shown their body produces plenty of lactase can still experience symptoms after consuming a significant amount of lactose in one sitting.[89] Finally, the fact that genes contribute to adult lactase production does not mean that it cannot change. In fact, a body's production of lactase does adjust to environmental stimuli.

Nevertheless, one need not go far to find cruder readings of what variable lactose digestion signifies. In recent years, White supremacists around the globe have embraced the concept of lactose intolerance as a distinguishing debility of non-White people. A map of the geographic distribution of lactose tolerance originally published in *Nature* has been copied and pasted into racist chat threads on the online discussion forum 4chan.[90] Internet trolls disrupting a livestreamed art installation in 2017 with Neo-Nazi chants also messily downed large jugs of milk as a symbol of White racial identity.[91] White nationalists then added milk-bottle emojis to their online profiles.[92] Nor is this sort of retrograde rot confined to professed racists and online trolls. A 2014 article in the *Journal of Economic Growth* suggested that the frequency of lactase persistence in a population in 1500 AD correlated strongly with that population's "level of economic development in the precolonial era."[93] *The Economist* chimed in a few months later in an article titled "No Use Crying: The Ability to Digest Milk May Explain How Europe Got Rich."[94] We have not traveled far, it seems, from those nineteenth-century claims that milk biologically equipped Europeans for conquest. Today they are couched in more opaque jargon, but their essence is the same.

Framing differential lactose digestion as a race-associated defect has fed demonstrations and arguments like these. Surely the scientists who have framed it this way did not intend to support crass prejudice. Still, it is not hard to imagine how an anodyne description of lactose intolerance like this one in the 2006 *Oxford Dictionary of Biochemistry and Molecular Biology* provides succor to someone already convinced that Chinese people are inferior: "This condition is particularly prevalent in the Orient."[95]

In the early twentieth century, a relatively new feature of Euro-American diets—namely, the prodigious consumption of cow's milk in fresh, liquid form—came to be revered as a key to health and national success worldwide. Chinese nutrition scientists saw the relative rarity of this habit in their country as one explanation for the country's political and economic subjection to milk drinkers. Over the course of the twentieth century, accordingly, both the People's Republic of China and Taiwan worked hard to expand their respective dairy industries and increase milk consumption. By the early 1960s, thanks in part to the large amounts of milk that the American government was then dumping into bodies unaccustomed to it, astute observers became aware that milk metabolism varies from one person to another. The variation more characteristic of people of northern European descent—lactase persistence—came to be seen as the norm or ideal, and the much more common variation came to be seen as "lactose intolerance," a deficiency.

The story of lactose intolerance shows that nutritional imperialism did not entirely recede in the postwar world, even as former colonies sought and won their independence and former colonial powers began to explicitly disavow imperialism. In much the same way that modernizers ignored China's dairy tradition, scientists later came to see Chinese people as constitutionally unable to process dairy. Early twentieth-century narratives about how this had disadvantaged them over time accordingly reemerged.

Milk is not the tonic that saves nations, and milk has never been inimical to Chinese cuisine, culture, or bodies. Yes, lactose digestion does differ from person to person and the human spectrum of lactose digestion does follow a rough geographic pattern. But medicalizing differences in lactose digestion has done more to reproduce the narrative of Chinese inferiority that we inherited from imperialists than it has done to improve public health. It is past time, therefore, to stop doing it. Here's a radical suggestion. Those who enjoy milk in any of its diverse forms can continue to do so regardless of their race and nationality. Those who do not enjoy it can happily consume a milk-free diet that promotes health and growth; they need not feel that their bodies are defective, despite the rhetoric coming from medical and dairy-industry journals and from unrepentant racists.

The next chapter will explore a final example of nutritional imperialism that has emerged even more recently than the concept of lactose intolerance: the invention of so-called aldehyde dehydrogenase deficiency, sometimes colloquially called "Asian Flush." Just as lactose intolerance has drawn on and fed popular notions about a Chinese fear and loathing of milk, aldehyde dehydrogenase deficiency has reinforced old stereotypes about Chinese people being temperate drinkers more prone to sedatives than to stimulants. Unpacking the social, political, and economic circumstances around the invention of this deficiency will help us understand its similarities with the earlier examples we have seen.

5

Temperance as Sickness

Alcohol Intolerance and "Asian Flush"

Chinese people, according to some modern observers, have an alcohol problem. It is not the problem you may be expecting after reading those words: It is not high rates of alcohol dependency or addiction. It is, instead, an inability to metabolize large amounts of alcohol quickly, caused by distinctive versions of the genes responsible for making ethanol-digesting enzymes. In everyday life, this supposed problem goes by the name "Asian Flush," or the nicer-sounding "Asian Glow." Its clinical names include "aldehyde dehydrogenase deficiency" and "alcohol metabolism deficiency."

A person who suffers from this so-called disorder rapidly flushes a bright red or purple when drinking alcohol and can experience from a relatively small amount of alcohol the rapid heartbeat, nausea, and headaches that others experience only after drinking much more. Besides these uncomfortable symptoms, it may also increase one's risk of oropharyngeal, esophageal and gastric cancer, and hepatitis—at least, that's what scientists and entrepreneurs invested in curing Asian Flush tell us. Some have even linked it with heart disease, neurodegenerative disorders, diabetes, and stroke.

At this point in the book, it should not surprise the reader to learn that I do not consider Asian Flush a deficiency. When placed in the context of all of the previous examples this book has explored, aldehyde dehydrogenase deficiency looks like the latest instance of nutritional imperialism. Like the earlier examples, it has roots in a late-nineteenth-century observation that Europeans and Americans made about how Chinese people were supposedly unlike themselves. In this case, the observation was that the Chinese were immune to drunkenness and alcoholism. This perception rested, like earlier perceptions of a meat-free and milk-free China, on a myopic understanding of history that projected contemporary circumstances back into the ancient past. And while some observers admired what they understood to be Chinese temperance, not all did. Some argued that alcohol, consumed appropriately, was a habit that equipped one for urban life in an industrializing world and therefore for modernity. Drinking alcohol supposedly gave a person the energy and stamina necessary to confront a newly fast-paced society, and those who wouldn't or couldn't drink it were only cementing their lowly place in the geopolitical order.

As in the case of other icons of modern diet—wheat, meat, and milk—the consumption of alcohol increased in China over the course of the twentieth century into the present, to the benefit of both domestic and foreign producers. And as in the case of milk, scientists discovered in the late twentieth century a physiological obstacle that was prevalent among Chinese people. This obstacle was subsequently reframed as a deficiency. By now it has become a "hereditary disorder." Thus, this latest example exhibits all the features of nutritional imperialism: perceiving turn-of-the-twentieth-century Western and European habits on the one hand and Chinese habits on the other as timeless characteristics of incommensurable civilizations and perceiving a Chinese difference in consumption and digestion as a disease. In fact, this example is more egregious than the other cases we have analyzed. It shows how a difference that might, under different assumptions, have been interpreted as an advantage—a genetic difference that provides some protection against alcoholism—instead came to be seen as a shortcoming. Below, we will explore how scientists came to invent this latest deficiency.

Selectively Seeing Temperance in Chinese History

Long before scientists began to study alcohol-induced flushing and unravel its genetic basis, European and American observers had begun to believe that Chinese people drank relatively little. As the historian Li Shang-Jen has shown, in the late 1800s, after several decades in China, the Scottish physician John Dudgeon wrote admiringly of Chinese drinking habits, noting that although strong liquor was cheap and readily available, the Chinese did not drink in excess or get drunk, in contrast to the European poor.[1] The American medical missionary John Glasgow Kerr concurred, writing, "The fact that drunkenness is not a common vice of the Chinese must have been noticed by most persons who have had much intercourse with the people."[2] Nearly 50 years later William Hector Park, an American missionary who had been living in Suzhou for more than 30 years, reflected that "as a boy at school I read in my geography that opium was more harmful to the Chinese than alcohol was to the people of other lands, and the impression made on my mind was that while Chinese used opium they did not use alcohol." The conventional wisdom, Park implied, was that the Chinese "are now not much troubled with diseased conditions due to alcohol."[3] A Chicago theologian agreed that "China has never been cursed by alcoholism."[4]

Such statements deserve the same kind of critical interrogation that we have applied to claims about a traditional Chinese aversion to or ignorance of meat and milk. In fact, primary sources attest that Chinese people used and enjoyed large amounts of alcohol historically and even struggled with dependence; the perception of categorical difference had more to do with the startling rise in alcohol consumption in Western countries starting in the 1800s. Let us examine some of the evidence.

Alcohol has a long history in China. How long? It features prominently in some of the oldest artifacts that survive from the region. The archaeologist Patrick McGovern has analyzed the residue of a fermented rice, honey, and fruit drink in 9,000-year-old pottery shards at a neolithic village in north China called Jiahu 賈湖.[5] A significant proportion—at some sites, as many as half—of the excavated bronze ritual vessels from the Shang and Zhou period, now scattered in museums around the world, were forged to contain alcoholic drinks.[6] These were first offered to the gods and ancestors and then enjoyed by living participants in the ceremony. Later, in the tenth to eighth centuries BC, many poems recorded in the *Book of Odes* (*Shi jing* 詩經) extolled the pleasures and benefits of drinking. In short, alcohol seems to have been fundamental to ritual culture, to social connection, and to political life in Chinese states from a very early period.

From an early period, too, Chinese observers expressed ambivalence about alcohol's effects. As Roel Sterckx has shown, Chinese governments up to the late third century AD justified state control of alcohol production by emphasizing the moral and political consequences of unbridled alcohol use.[7] Some early histories blamed insobriety for the decline and fall of the Shang dynasty in the eleventh century BC. They noted that Duke Wen of the succeeding Zhou dynasty officially admonished officials of the new regime not to drink, lest their state suffer the same fate. While we cannot accept the charge at face value—such moralizing was a typical way that scholars serving a new regime made the old one look bad—it highlights a shared concern among the Chinese elite that alcohol use could corrupt virtue and disrupt social order. Moreover, in practical terms alcohol production required grain that might otherwise be used as food, so if the well-to-do had too much to drink, the poor might have too little to eat.[8]

But there was a paradox: Because of its importance in state ritual, a government could not just ban alcohol altogether. Even Duke Wen made an exception for beverages used in sacrifices to the spirits and ancestors. The sign of virtue when it came to alcohol, then, was not total abstinence but using it responsibly in a ritual context: first offering a drink to the spirits and ancestors, then to the elders, and not consuming it in such amounts or with such speed that one became drunk and disorderly. Poems describe idealized banquets and sacrifices in which people are continuously drinking but doing so in a decorous way and not becoming silly or insensible. Confucius himself is said not to have limited his drink but never to have become drunk either.[9]

In later centuries, outside of the context of state and community ritual, scholars and poets celebrated *jiu* 酒—sometimes translated as "wine," though in the early and medieval periods it referred to drinks made from grain, so I will use "beer" or "ale" here. They saw it as freeing the drinker from conventional social expectations and official obligations. The Seven Worthies of the Bamboo Grove, third-century-AD literati who rejected the ideological order behind the recently collapsed Han state, presented drinking as a way to escape political danger. Being too drunk to be recruited into official service or to be blamed for schemes gone awry could help a person survive a period of deadly political intrigue. Boozing could also, texts about the Seven Worthies suggest, help one reach great heights of creative accomplishment.[10] By the eighth century, when the poet Li Bai wrote his celebrated ale-soaked verses, a perceived connection between alcohol and creative brilliance was firmly entrenched.[11]

At the same time, doctors and others recognized alcohol's potential to damage health. The drug formulary *Essential Formulas [Worth] a Thousand in Gold* (*Qian jin yao fang* 千金要方) by the seventh-century-AD doctor Sun Simiao enumerated drinking's health consequences in his elite world. The drink-caused disorders that the text described ranged from the immediate and merely uncomfortable (e.g. headache, bloated belly, and restlessness or trouble sleeping) to the more serious and long-term (ale poisoning, rotting the five *zang* 臟 organs, depleting one's *qi* 氣, and developing ulcers in the throat and on the tongue), as well as the generic "alcohol sickness" (*jiu bing* 酒病).[12] Some drinkers evidently wished to stop but couldn't: The largest number of formulas in this section aimed to help a person "quit alcohol" (*duan jiu* 斷酒).

Among those trying to quit, perhaps surprisingly, were the hard-drinking literati who celebrated alcohol as a social lubricant that spurred creativity and liberated the drinker from convention.[13] The fourth-century AD poet Tao Yuanming 陶淵明's piece entitled "Stopping Ale" (*Zhi jiu* 止酒) is an early but not unique example of poems on this same theme:

> All my life I've never stopped drinking;
> When I stopped my feelings knew no joy.
> If I stopped in the evening I could not sleep;
> If I stopped in the morning I could not get up.
> Every day I wanted to stop,
> But if I stopped my pulse and circulation turned erratic.[14]

Tao's poem was only an early example of a whole category of "giving up ale" poems. There are dozens of poems on the topic, especially in the middle period. Here is another example, an excerpt from a poem by Mei Yaochen 梅堯臣 called "Judge Fan Implores Me to Stop Drinking" (*Fan tuiguan quan yu zhijiu* 樊推官勸予止酒):

> Today, though my teeth and hair are falling out,
> I love to drink, but I can't drink as much,
> Because every time I drink I vomit and have diarrhea.
> How could my six organs be in harmony?
> I'm hungover in the morning and I can't lift my head.
> The room spinning like a whirlpool.
> I search for happiness but get sickness.[15]

Edwin van Bibber-Orr, a literary scholar who surveyed Song dynasty poems on alcohol-induced sickness, argues that the poems show the poets suffered from what the *Diagnostic and Statistical Manual of Mental Disorders* calls "alcohol use disorder."[16] Like many historians of medicine, I reject the premise that one can meaningfully diagnose premodern people using current disease definitions.[17] What this literature does show us, however, is that many

Chinese scholars experienced alcohol as a source of suffering, not just a source of liberation and social connection—and that, in turn, calls into question nineteenth-century missionaries' assertions that Chinese people did not struggle with drink.

The visibility of Buddhism and its moral injunctions against alcohol influenced modern perceptions that Chinese society tended toward teetotalism, just as Chinese Buddhist prohibitions of meat contributed to China's early twentieth-century image as a "vegetarian nation," as chapter 3 discussed. It is true that the rising importance of Buddhism in the centuries following its introduction to China around the first century AD increased the moral disapprobation surrounding alcohol. Many Buddhist writers objected to the way it muddled the mind and prevented one from attaining enlightenment, not to mention the way it decreased inhibition and facilitated immoral actions such as violence or illicit sex. Some presented drinking as a sin whether or not immoral behavior ensued. Even facilitating others' drinking without consuming alcohol oneself could incur a severe karmic penalty. Some scriptures said that a person who passed a cup of ale to another would be reborn without hands for 500 years. Buddhist clergy, who occupied the same elite circles as the literati officials we have discussed so far, increasingly turned to tea as a morally acceptable drink that would allow them to participate in social occasions without losing mental acuity or arousing the passions. Still, not all Buddhist texts prohibited alcohol, and evidence suggests that in some times and places Chinese Buddhist monks indulged, sometimes a lot.[18] Buddhists also did not generally claim that alcohol was bad for one's health. In fact, some Buddhist texts suggested that it might be morally acceptable for a sick person to take alcohol medicinally as long as they gave it up once cured.[19]

The medicinal exception hints that alcohol's perceived relationship to health was as complex as its perceived relationship to moral and social order. Although, as we have seen, doctors and drinkers were alert to health problems that drinking caused, they also believed that alcohol could heal as well as harm. This resonates with the belief in classical Chinese medicine that potent substances can function as either poison or drug.[20] An old form of the Chinese character for alcohol, 酉, is even embedded in the character for "medicine" or "healer," 醫, constituting its bottom half. Medicinal grain-based drinks imbued with herbs were valued as tools to promote health and longevity, as they remain in Chinese communities today. Therapeutically, alcohol—sometimes administered orally, sometimes topically—featured in formulas meant to treat conditions associated with cold and dampness or stagnation of the blood. According to books of *materia medica*, alcohol warmed the body, stimulated the *qi* and blood, and supported the transformative activity of the spleen and stomach. Drug formularies indicated that alcohol, when used to compound herbs, could increase their potency and efficacy.[21] Not surprisingly given this therapeutic rationale, medicinal liquors have most typically been used by the elderly and in the winter, a life stage and a season when cold and depletion disorders are thought to be common.[22]

But what about one particular negative health consequence of drinking, the one at the center of this chapter—namely, "Asian Flush"? A reader might like to know whether there was any precursor in premodern China to this modern scientific concept. After all, if it is true that a significant minority of people in China today have the genetic variation that causes a dramatic flushing reaction in response to alcohol, presumably their ancestors also had it in the past. One might be curious, therefore, to know whether Chinese medical writers observed differences in the way individuals or groups tolerated alcohol.

The best short answer to such questions is that Asian Flush did not exist before the 1990s, when the term was coined. With a less-strict definition that does not require the name to exist, one might date the supposed condition back to the 1970s, when scientists began to write about a type of alcohol metabolism distinctive to Asian experimental subjects. But from the perspective of a medical historian, it certainly did not exist in premodern China.

Let me explain what I mean by this. I do not mean to suggest that no one in premodern China had the genetic variation that causes many people today to flush red when drinking alcohol. Presumably many people did, and presumably it caused the same sort of physiological response then that it does today. We do have evidence that people in premodern China observed differences in the way individuals responded to alcohol. For example, the tenth-century scholar Su Shi 蘇軾 indicated that he could not drink as much as others and that even a small amount of drink sank deep and left him feeling half-drunk.[23] Northerners were thought to have a greater tolerance and affinity for alcohol than southerners. One explanation texts offered for this was that ale, a warming substance, helped northerners survive and endure winters much colder than the ones in the south.[24] To return to Sun Simiao's *Essential Formulas [Worth] a Thousand in Gold*, discussed above, it is suggestive that the remedies targeted not only consequences of chronic drinking but also some men's inability to drink a socially expected amount. Among the formulas in Sun's collection lurked a few meant to correct an *aversion* to alcohol or an excessive sensitivity to its effects. If you wished to play drinking games without getting drunk, for example, or overcome a distaste for ale, *Essential Formulas* had a recipe for you. Some people apparently did experience being unable to drink properly as a kind of disability in itself and something worth addressing medicinally.

In other words, yes, it seems that some Chinese people before the twentieth century flushed or felt sicker, faster compared with their friends when they were drinking. To acknowledge this is not to argue that Asian Flush existed hundreds of years ago, however. No disease, disorder, or deficiency can be satisfactorily reduced to the mechanics of molecules. Turning a sequence of chemical reactions into a *diagnosis*—a deficiency, for example—requires cultural and social assumptions. Deficient compared to what or whom? People in, say, the seventh century did not live in the same racialized world that we live in today; Chinese doctors of that time had no concept of "Asia," a homogenizing fiction that obscures the enormous diversity of the people living in the region it applies to. They could not, by definition, have had a concept of Asian Flush.

Just as important, the concept of a disorder inherent to a large group of people sits uneasily with the individualistic orientation that has often characterized classical Chinese medicine. A doctor was likely to view one person's reaction to drinking, like everything else about their state of health, as unique to them.[25] Since in Chinese medicine alcohol's function was to warm and dry, a medical view of someone who flushed violently or felt ill after ingesting alcohol might be that the person's constitution tended toward heat and that the alcohol had exacerbated that inclination. A reasonable response would be for that individual to abstain from alcohol and consume cooling foodstuffs (dairy, perhaps).

Alcohol's historical significance in China was complex. As an important element of ritual, it supported the social and moral order even while, as a powerful intoxicant, it threatened to undermine both. It could either cement or unravel social relationships. And it could sicken or strengthen a person, kill or heal. Asian Flush, however, did not exist before the necessary social, political, and cultural circumstances emerged to turn a sequence of chemical changes into a race-related disability—and that did not happen until the last decades of the twentieth century, as subsequent sections will show. The next one will explore the early stage in the development of those circumstances, as well as the perceptions and provision of alcohol in China in the late 1800s and early 1900s.

Alcohol as Agent and Symbol of Modern Imperialism

Alcohol in China was both a symbol and an agent of imperialism. This may be a surprising claim. When one thinks about the history of Western imperialism in China, alcohol is not the first drug that comes to mind. That is opium, and for good reason. The armed conflicts that secured British and other European states' privileges in nineteenth-century China were called the "Opium Wars," after all, because Chinese resistance to Europeans peddling the narcotic was a precipitating cause. But alcohol is the drug that better represents imperialism in China *after* the nineteenth century. More stringent opium control in China and in Japanese-ruled Chinese communities meant that for foreigners, selling opium to Chinese users in the early twentieth century was harder than it had been in the past.[26] The public consensus, moreover, was that recreational opium use in any amount was an evil to be eliminated.

Alcohol was different. While there was plenty of moralizing around alcohol production, sale, and use, beer and wine were nevertheless profitable and growing aspects of the European presence in China. The best-known example is the Anglo-German Brewery Company in the port city of Qingdao, founded in 1903, whose story Zhiguo Yang has expertly related.[27] German gunboats had forced the Qing government to cede Qingdao to Germany at the end of the nineteenth century, and within a few years German and British businessmen had established a brewery there to serve the growing number of Europeans living in China's treaty ports. In other words, the same semi-colonial context that had introduced Western styles of milk production and consumption, discussed in chapter 4, also introduced Western-style beer and breweries.

The Anglo-German Brewery Company's status as an emblem of imperialism did not end when Germany lost its Chinese territory during the First World War. Japan claimed the German concessions, so Japanese businessmen came to own and operate the company. They expanded the market for its products to include Chinese customers, doubling production capacity by the 1940s. When Japan, in its turn, lost a war and thus its hold on Chinese territory, Chinese businessmen and in short order the Chinese Communist Party came to control the brewery. Even then it continued to represent a kind of dependence on foreign powers, however. Until the 1980s, Qingdao Brewery sold more to overseas customers, particularly in Hong Kong and Macao, than to domestic ones; the government of the People's Republic of China relied on the brewery to bring in foreign currency.[28] Although, as Yang has argued, the Tsingtao (Qingdao) brand has come to represent China in the global beer market and appeals to Chinese nationalism helped ensure the company's survival and success after World War II, its longer history as a symbol of imperialist domination lurks ironically beneath that triumphant surface.

Tsingtao Beer may be the most familiar example of alcohol in the service of imperialism, but it is hardly the only one. In French Indochina, the colonial government's alcohol monopoly was one of the most profitable and hated institutions of its rule.[29] As prohibition movements took effect in North America in the early decades of the twentieth century, US and Canadian brewers began looking abroad to make up for their lost markets at home. China's large population attracted their attention just as it attracted other groups of evangelists and entrepreneurs. The perception that Chinese drinkers were singularly immune to alcohol addiction gave cover to the liquor and beer promoters, muting the kind of moral opprobrium by then attached to selling opium. Some brewers even advertised alcohol as a "sure cure for opium."[30]

Temperance activists decried this as shameless opportunism and questioned the assumption that alcohol would harm Chinese people less than it harmed Europeans or Americans. Park, the missionary who had been taught at school that opium was China's drug of choice, fumed that contrary to the common perception, "Alcohol has the same effects on the people of China it has on any and all other peoples of the earth who use it. . . . Foreign liquors are being introduced into China and if foreigners find as much money can be made out of their sale as used to be made out of the sale of opium they also will be found peculiarly adapted to the Chinese constitution."[31] The author of an article titled "Shall We Make the Chinese Drunkards?" similarly averred, "What is dangerous to America is certain to be dangerous elsewhere."[32] At the Paris Peace Conference in 1919, anti-alcohol activists urged participants in the nascent League of Nations not to allow the winning countries to "alcoholize" China by exporting drink there in large amounts—

to no avail, apparently. By 1924 the amount of foreign alcohol imported to China was double what it had been 10 years earlier.[33]

Other Western writers residing in China who were less concerned about the moral stakes of drinking likewise took a jab at the conventional wisdom. Writing for the newspaper *China Weekly Review*—the first one founded by professional American journalists in China—Tom Doyle observed, "Lots of Chinese like to get drunk in exactly the same way that lots of other people like to, the only difference being that the Chinese usually drinks the other fellow completely under the weather."[34] The main reason they did not drink quite as much or as often as his fellow Americans, he declared, had nothing to do with taste or culture or religion. "Just one reason occurs to explain why China is not as drunk as any other civilized and progressive race," he wrote, "and that reason is purely economic." If Chinese consumers could afford more alcohol, Doyle had no doubt that they would be drinking more. He scoffed, "Sober China—sober because she is broke!"[35]

Doyle's quip hints at a context that resembles what we have seen earlier in the cases of milk and meat and a misreading of a historical moment similar to what we saw in those cases. Just as the largest impediment to increasing consumption levels of those relatively expensive products was poverty, so it may have been for alcohol as well. The misreading was to take that peculiar moment as representative of all of history. Chinese consumers had not been so poor as compared with European and American ones until relatively recently. As Kenneth Pomeranz has shown, in the eighteenth century the residents of the Yangzi River basin, an area and population comparable to that of Great Britain, had been as well-nourished and long-lived as their British contemporaries. More to the point, they spent no more of their income on food and consumed non-essentials like sugar and tea in similar per-capita quantities.[36] As the Chinese population grew steadily more impoverished over the course of the nineteenth century, Europeans and Americans, by comparison, grew richer.

They also began to drink more, and differently, than they had in the past. As historians have chronicled, by 1820 American alcohol consumption had largely moved out of the home and into public spaces.[37] And by the turn of the twentieth century Americans were "drenched in beer," as Daniel Okrent has put it, consuming 24 times as much in 1890 as in 1850, despite a mere tripling of the population.[38] What made the Chinese look to Westerners like teetotalers in the 1920s, in other words, was a recent spike in visible consumption in the West more than any inherent Oriental reserve. The abiding American preoccupation with alcohol, and its regulation and resistance to regulation during Prohibition, can only have enhanced the sense that Americans had a problem with drink that others did not. Interpreting this peculiar historical juncture as a reflection of timeless differences between Chinese and Western civilization is what allowed the Chinese to become Westerners' teetotaling foil.

In addition, the idea that Chinese people shied away from alcohol fit the prevailing stereotype that they were passive and conservative. In the early twentieth century, a national fondness for alcohol was not always, in all quarters, cause for shame or alarm. Norman Smith has argued that in early 1930s Manchuria under Japanese rule, alcohol use was initially promoted as a "marker of modernity" and moreover healthy: It stimulated the circulation, kept a person warm and gave them energy, and fueled healthy socialization.[39] The historian Jeffrey M. Pilcher notes that in East Asian markets, beer in particular was associated with the "supposed robustness of European imperialists."[40] The marketing specialists attempting to get more Chinese consumers to purchase Tsingtao Beer in the late 1940s emphasized its health-giving qualities, as Zhiguo Yang has reported. They observed that it could "sharpen your sense of smell and stimulate your appetite for food" and that the body could easily absorb the "chemical elements" that it contained, which added up to its being "a highly nutritious beverage." Their boss, Zhu Mei 朱梅, the first Chinese director of the brewery, argued that their product was "of great importance to the health of the Chinese nation."[41] Thus, a taste for the right kind of alcohol, consumed in the right settings, was presented as a boon to national health and an index of national modernity.

Alcohol, then, was the craving of ambitious and progressive nations. In contrast, opium was perceived as the characteristic Chinese vice. This was partly for historical reasons: The prominent role of opium in the disastrous nineteenth-century Qing wars against Britain associated the drug with Chinese military vulnerability and political weakness. But the distinction also drew on stereotypes about Chinese quiescence that chapter 1 explored in the context of basal-metabolism research. Opium was a sedative, alcohol a stimulant. As Smith has pointed out, a drug such as opium that one consumed while lying down and that numbed a person and slowed their heart rate and breathing was the kind of drug that a person might expect Chinese people to appreciate if they believed that the Chinese were innately suited to "indoor activities like writing poetry." Europeans, by contrast, "seemed to favour alcohol," a preference that some correlated with their "excel[ling] at outdoor activities, including driving cars, boats, and planes."[42]

Considering how anecdotal was the evidence that undergirded Chinese people's global reputation for teetotalism, and considering evidence contradicting that reputation, preexisting assumptions about Chinese passivity and weakness surely contributed to it. While opium use was indeed a problem in early twentieth-century Chinese communities, alcohol use was just as common.[43] Even a casual perusal of the literature on historical Chinese drinking games, catalogued by the hundreds and designed, in many cases, to maximize drinking companions' intake, is enough to make a reader doubt any categorical claim about Chinese temperance—as is a reading of any late-imperial Chinese novel, full of boozy encounters of one kind and another.[44] And the contrast with perceptions of Japan is instructive. Although the gene that causes flushing and can discourage alcohol consumption appears as frequently among Japanese as among Chinese people, early twentieth-century Japanese observers saw their own compatriots as more alcohol-inclined than opium-inclined. Just like Europeans, they were quick to point out, they were an "active" race of drinkers rather than a "lethargic" one of opium smokers like the Chinese.[45]

To summarize, the impression that Chinese people were unusually abstinent took shape at the height of the age of imperialism and reflected the economic and social circumstances of that moment as well as prevailing stereotypes about Chinese character. That did not prevent the idea from sticking around to inform a program of scientific research in the second half of the twentieth century. That is the story we will explore in the next section of this chapter.

The Science of Alcohol Metabolism and the Making of a New Deficiency

By the late twentieth century, scientific research into alcohol metabolism had provided a genetic explanation for the old belief that Chinese people had a different relationship with alcohol than did others. In what is by now a familiar move, scientists also reframed that difference as a sickness: a genetic deficiency that put one at higher risk of deadly diseases.

To understand that change historically, we must begin with a sketch of how scientists came to understand alcohol's effects on the human body in the modern period. Starting in the 1910s, researchers in the United States and Europe were developing novel techniques to assess alcohol metabolism. Concerns about public order and public health motivated the earliest research in this area. On the public-order side, police and courts sought reliable and convenient ways to measure an individual's degree of drunkenness so that they could determine culpability for various kinds of offences. In 1914 the Swedish chemist Erik M. P. Widmark devised a method for measuring alcohol concentration in the blood that was subsequently widely used in Europe.[46] In the 1930s it was supplanted by newer tests that could divine blood-alcohol concentration by sampling breath instead of blood, a more convenient and less invasive method. An improved version of the breath tests known as the Breathalyzer appeared in the 1950s, followed by a handheld electronic Breathalyzer in the 1970s.[47]

Meanwhile, physicians had been seeking pharmaceutical tools to combat addiction. In the late 1940s they found one in the form of disulfiram, a chemical compound invented in the 1880s, which had previously been used in rubber vulcanization.[48] A couple of Danish scientists were exploring its potential as a treatment for parasitic worms when, by administering it to themselves, they observed an odd side effect that some earlier rubber workers had also experienced: Exposure to disulfiram induced an extreme and uncomfortable reaction to drinking alcohol. Those unfortunate enough to experience the combination of these two substances found their skin flushing hot and red, heads aching, and stomachs wracked with nausea. The scientists initially perceived this as a setback, since it made disulfiram so undesirable as a deworming agent, but after a chance encounter with a physician colleague who specialized in treating alcoholism, they realized they could reframe it as a therapeutic feature.[49] It could be used to make alcohol repulsive to addicts.

As scientists began to develop disulfiram (later known by the brand name Antabuse) into an anti-addiction drug, they conducted research to elucidate its mechanics. How did disulfiram so dramatically transform the experience of drinking alcohol? This research helped scientists begin to understand more precisely the chemistry and physiology of alcohol metabolism.[50] Over time, the basic outlines began to emerge. First, in the stomach and the liver, an enzyme called alcohol dehydrogenase (ADH) converts ethanol into acetaldehyde. A second enzyme, aldehyde dehydrogenase (ALDH), then converts acetaldehyde into acetic acid. Acetic acid breaks down into carbon dioxide and water, which the body excretes (Figure 5.1).

Steps in alcohol digestion:

STEP 1. Ethanol + alcohol dehydrogenase (ADH) → **acetaldehyde***

STEP 2. **Acetaldehyde** + aldehyde dehydrogenase (ALDH) → acetic acid

STEP 3. Acetic acid → Carbon dioxide and water

*Acetaldehyde (in bold) is the dangerous compound

Figure 5.1 Steps in alcohol digestion. Image by author.

Acetaldehyde, the compound at the midpoint of this process, is toxic and causes a number of uncomfortable sensations such as flushed skin, headaches, and nausea. But for most people the ethanol-to-acetaldehyde conversion is slower than the acetaldehyde-to-acetic acid conversion, so acetaldehyde becomes harmless acetic acid before it can cause any perceptible symptoms. Taking disulfiram changes that, because disulfiram competes with acetaldehyde for ALDH, meaning that less of the enzyme is available to process acetaldehyde. Thus, acetaldehyde builds up and causes the unpleasant sensations just described.

Medical and law-enforcement goals converged in the 1960s in research meant to determine whether different bodies processed ethanol at different rates, and if so, what factors affected the course of digestion. It was common knowledge that some people maintained an air of sobriety after consuming quantities of alcohol that incapacitated others and that some people seemed to sober up more quickly than others. Understanding why might help police assess a particular individual's incapacity more accurately, and it might help doctors understand how different patients would experience disulfiram. Did the differences reflect gender, body mass, or the amount of alcohol a person habitually consumed? And did they, scientists soon asked, reflect race?

The earliest race-inflected studies of alcohol metabolism appeared in medical and scientific journals in the early 1970s. In 1971, Canadian researchers at the University of Alberta published a study comparing the rates at which blood alcohol concentration declined in "21 Eskimo, 26 Indian and 17 white males" after an intravenous infusion of alcohol. They sought to determine whether "unverified medical and law-enforcement reports stating that Indians jailed while intoxicated take much longer to sober up than whites" held up to scientific investigation.[51] According to this study, they did. The researchers concluded that "the natives metabolize alcohol at a significantly slower rate than the whites," and having dismissed some environmental factors that they had thought might play a role, they concluded the contrast was "probably due to genetic differences which may, in turn, be due to differences in diet over many generations."[52] In subsequent years, other studies challenged this conclusion, finding no difference in metabolism rates between Native and White subjects or finding faster rates in Native than in White subjects.[53]

When it came to Chinese drinkers, the "unverified" conventional wisdom was that Chinese and other East Asian people—in the reports of the 1970s generally labeled "Oriental" or "Mongoloid"—seemed less prone to alcoholism than Europeans and Americans. This was a continuation of the refrain that had emerged a century or more earlier. As the historian Sarah Tracy has highlighted, what researchers viewed as alcoholism changed over the course of the 1900s.[54] Nevertheless, the conclusion that Chinese drinkers suffered it less often than others survived the changes. In 1946 the American anthropologist Weston La Barre, drawing on his recent two-year residence in China as a member of the Sino-American Cooperative Organization, published "Some Observations on Character Structure" among the Chinese. Although La Barre was neither a medical doctor nor a China specialist, his observation that "the Chinese are not so violently addicted to excessive use of alcohol as have been some North European peoples" was frequently quoted by later researchers, many of whom agreed with him that "the solitary Chinese drinker is unthinkable and the problem alcoholic an evident rarity."[55] Other anthropologists concluded that alcoholism was notably rare in Chinese immigrant communities in the United States.[56]

Some Chinese observers shared this belief as well. In 1953 a researcher at National Taiwan University's School of Medicine oversaw an intensive survey of nearly 20,000 Chinese people in Taiwan that concluded, "Although the use of alcohol is as old as the history of China, and is fairly extensive, alcoholism does not constitute a problem in Chinese communities." Moreover, in Taiwan, "The rate of alcoholism was quite low compared to that of Western civilizations."[57] Such observations continued in subsequent decades.[58]

Before the 1970s, researchers had attributed the presumed lower incidence of alcoholism in China and the Chinese diaspora to cultural and social factors. For example, the eminent psychiatrist Morris Chafetz, who a few years later would become the first director of the US National Institute on

Alcohol Abuse and Alcoholism, wrote in 1964 that the "strong moral code of the Chinese, derived from Confucius," prevented Chinese people from getting drunk.[59] A 1968 study of alcoholism in New York's Chinatown conducted by Richard P. Wang for Cornell University's medical school pointed toward Daoism and Confucianism as ideologies that led people toward enjoying things such as alcohol in moderation. It also credited the high premium on "rational control of behavior through intellect" that Wang claimed characterized Chinese culture.[60] Not all those seeking an explanation resorted to Confucius. Some observed that drinking culture in Chinese communities was consistently social; people drank at boisterous banquets and in the company of friends, lovers, and associates. That was the point of La Barre's comment that a solitary Chinese drinker was "unthinkable."

In the 1970s, however, a new explanation emerged for the temperance that researchers thought characterized the Chinese. In an article about "Ethnic Differences in Alcohol Sensitivity" published in *Science* in 1972, Peter H. Wolff, a research psychiatrist at Boston's Children's Hospital, opened with a premise that resembled the one earlier observers had expressed: "Mongoloid" people, he asserted, had a "lower incidence of alcoholism" compared to "Caucasoid groups."[61] But instead of looking to culture to explain this observed difference, Wolff looked to heredity. He examined 171 subjects, some of them young adults and others infants, from the United States, Japan, Taiwan, and Korea, had them ingest a measured dose of alcohol, and then, having assessed whether or not their faces flushed, measured the degree of flush by applying an optical densitometer to their earlobes. He found that many of the Japanese, Taiwanese, and Korean subjects experienced "a marked facial flushing and mild to moderate symptoms of intoxication" after ingesting alcohol, while the White subjects did not.[62] Since the same contrast appeared among the infants as well as the adults, Wolff felt confident that it reflected a genetic difference rather than a cultural or dietary one. His research blazed the trail for what would eventually become a well-trodden route of scientific inquiry. In the decades that followed, scholars around the world would contribute many more studies to elucidate how genes shape alcohol metabolism.

It is worth remarking how alien this experiment and those that followed were to any real-life experience of drinking. Subjects entered a lab one by one and absorbed alcohol in a carefully measured amount. Sometimes this took the form of beer or another commercially available drink that a person might willingly consume outside of a laboratory environment, but more often it was pure ethanol mixed with water or juice to a desired strength, drunk from a lab cup under the supervision of a technician. Occasionally it was injected directly into the subject's bloodstream. Then the technician tested their response, applying instruments to their earlobes and faces. The drinkers were not playing games or writing poems or commiserating with friends about the vicissitudes of life. They were not eating delectable food or appreciating the sweet fragrance of a warm cup of rice beer. In fact, experimental protocol in many studies mandated that the participants come in with an empty stomach. They were not offering libation to the spirits or ancestors to fulfill family and community obligations and to pursue harmony and prosperity. They were not engaged, in other words, in an activity as rich with meaning and emotion as most activities that involve drinking are. Anthropologists have shown that the context of medical treatment influences how well it works; presumably, the context in which one consumes alcohol can also influence its effect on the body.[63]

Before long scientists began to understand more precisely the chemical makeup and action of alcohol-metabolizing enzymes. By the end of the 1970s, they had zeroed in on acetaldehyde as the main cause of alcohol's skin-flushing effect. East Asian drinkers whose skin flushed deeply were experiencing something similar, researchers realized, to what patients who took disulfiram experienced when they drank. In both cases acetaldehyde accumulating in the bloodstream caused symptoms, but in East Asian drinkers it was not because they had ingested a drug that was siphoning off the acetaldehyde-digesting enzyme ALDH. Instead, their bodies produced less ALDH to begin with. More specifically, their bodies produced less ALDH2, the version of ALDH that converted acetaldehyde to acetic acid most readily.

One imaginable response to such a discovery might be celebration. Some bodies, it seemed, had a built-in resistance to alcohol akin to the one doctors used medicine to induce in addicted patients. Perhaps scientists could learn something from these bodies naturally equipped to reject a substance that was simultaneously attractive, addictive, and poisonous. They might thus make more progress preventing and treating alcoholism.

Instead, having a genome that resulted in low levels of ALDH2 quickly came to be labeled "deficient" and "abnormal." A multinational team of scholars led the way in studies of what they began calling "aldehyde dehydrogenase deficiency." In October of 1981 the Japanese scientist Shoji Harada joined forces with the Indian scholar Dharam P. Agarwal and the German geneticist H. W. Goedde at the University of Hamburg's Institute of Human Genetics to publish a brief research report in the *Lancet*, one of the world's most prestigious medical journals. In it, they described an experiment they had conducted with Japanese subjects whom they had divided into a "normal" and a "deficient" group based on whether or not their bodies had ALDH2. The "deficient" group was large—43% of the total pool of participants—and had much, much higher levels of acetaldehyde in their blood after an oral dose of alcohol. This, they concluded, caused "flushing syndrome in mongoloid peoples."[64]

Harada, Agarwal, and Goedde thus reframed the whole phenomenon in a way that had long-lasting consequences. They referred to flushing as a "syndrome" and to having a different configuration of ALDH isozymes as a "deficiency." What's more, they speculated that people who were "deficient" in ALDH2 in the liver were vulnerable to far more serious health problems than just temporarily flushed skin and discomfort. Such people, they concluded grimly, "may be at higher risk of acetaldehyde-related organ damage or of complications in utero resulting from maternal alcohol abuse."[65] On the other hand, the same team speculated in a follow-up article, this "deficiency" might have a positive effect in protecting deficient individuals from alcohol abuse and addiction.[66] This became the pattern in discussions of the genetic variation in ALDH production that was common among Chinese people: It was presented as both a biological explanation for low rates of alcoholism and as a genetic defect that exposed its carriers to severe long-term health consequences.

The idea that having few ALDH2 isozymes was a kind of defect and a potential medical problem lodged itself firmly in scientific journals. Figure 5.2, a table from a 1986 article by the Hamburg trio and their colleagues, estimates the "percent deficient" among study participants of different nationalities and races. A 1989 article in the journal *Alcohol* sharpened the rhetoric, describing bodies without ALDH2 as "deficient individuals" whose "isozyme abnormality" resulted from "a deranged acetaldehyde metabolism" with "toxic consequences."[67]

The earliest Chinese-language studies of this physiological variation came in the early and mid-1980s in the journal *Overseas Medicine* (*Guowai yixue* 國外醫學), a title indicating that the original inspiration for this kind of research in China came from abroad. The 1980s and 1990s saw a slow trickle

Frequency of ALDH Isozyme Deficiency in American Indians, Asian Mongoloids, and Other Populations

Population	Sample size	Percent deficient	Reference
South American Indians:			
Atacameños (Chile)	133	43	[6]
Mapuche (Chile)	64	41	[7]
Shuara (Ecuador)	99	42	[8]
North American Indians:			
Sioux (North Dakota)	90	5	Present study
Navajo (New Mexico)	56	2	Present study
Mexican Indians:			
Mestizo (Mexico City)	43	4	Present study
Asian Mongoloids:			
Japanese	184	44	[1, 3, 9]
Chinese:			
Mongolian	198	30	[10]
Zhuang	106	45	[10]
Han	120	50	[10]
Korean (Mandschu)	209	25	[10]
Chinese, living abroad	196	35	[8]
Koreans (South Korea)	75	27	[15]
Vietnamese	138	53	[8]
Indonesians	30	39	[8]
Thais (North)	110	8	[8]
Filippinos	110	13	Present study
Ainu	80	20	Present study
Other populations:			
Germans	300	0	[8]
Egyptians	260	0	[8]
Sudanese	40	0	[8]
Kenyans	23	0	[8]
Liberians	184	0	[8]
Turks	65	0	Present study
Fangs	37	0	Present study
Israeli	77	0	Present study
Hungarians	177	0	[16]
Matyo	106	0	[16]
Romai	84	0	[16]
Asian Indians	50	0	Present study

Figure 5.2 Frequency of ALDH Isozyme Deficiency in American Indians, Asian Mongoloids, and Other Populations. In H. W. Goedde et al., "Aldehyde Dehydrogenase Polymorphism in North American, South American, and Mexican Indian Populations," ***American Journal of Human Genetics*** **38, no. 3 (March 1986): 397.**

of Chinese articles on the topic, which widened into a durable stream starting in the early 2000s, with articles coming from institutions in all corners of China and addressing genotyping as well as various diseases for which variant ALDH genes are a risk factor.

In China and elsewhere, much of the relevant research during the genome-obsessed 1990s and 2000s strove to identify more precisely the genomic quirk responsible for this "deficiency." The culprit turned out to be a mutation in one allele (also known as a single-nucleotide polymorphism, or SNP) that came to be called ALDH2*2. The more common allele scientists dubbed ALDH2*1. The ALDH2*2 variant had first appeared 2,000 to 3,000 years ago in what later became southeast China and had persisted and spread rapidly across East Asia. Recent studies suggest that it now features in anywhere from 30% to 50% of genomes in China, Japan, the Koreas, and Vietnam.[68] Those two different alleles produced three different genotypes: Some people had two ALDH2*1 alleles, some people had two ALDH2*2 alleles, and some had one ALDH2*1 and one ALDH2*2 allele. The ALDH2*2 allele was associated with low levels of ALDH2 (or a total absence of it) and therefore with a reduced capacity to process poisonous acetaldehyde, so in East Asian drinkers, the ALDH2*2 allele caused the unpleasant and violent reactions that many experienced when drinking alcohol. ALDH2*2 turned out to be dominant, so it was not only those with a homozygous ALDH2*2/*2 genotype who flushed and felt sick; to a lesser degree, those who were heterozygous ALDH2*1/*2 did too.

By 2014, scientists were working to turn this genetic difference into a general pathology. A group of researchers at Stanford University and the University of Sao Paulo began to refer to having the ALDH2*2 allele as an "enzymopathy" associated with elevated risk for some of the most-feared diseases of our time: "cardiovascular diseases, diabetes, neurodegenerative diseases, stroke, and cancer."[69] One of the Stanford scholars, Che-Hong Chen, even established a research consortium specifically devoted to ALDH2 deficiency, calling it "the most common human enzymopathy in the world" and identifying Taiwan as a particular hotspot for the "aberrant gene" that caused it.[70] The chair of genetic medicine at Weill Cornell Medical College suggested that those who have the ALDH2*2 allele suffer from "one of the most common hereditary disorders in the world."[71] He and his colleagues proposed to use gene therapy to "correct" this disorder.[72] Research into the associations between ALDH2*2 and various kinds of chronic and fatal diseases has become a cottage industry.

Not coincidentally, there is money to be made in treating a "hereditary disorder" characteristic of 30% to 50% of the planet's most populous continent. In 2011 the Stanford scientists who have been so active in elevating the visibility of aldehyde dehydrogenase deficiency founded a pharmaceutical company called ALDEA, whose purpose was to develop drugs to treat this condition.[73] The company outlined a number of possible uses for the aldehyde dehydrogenase activators (ALDAs) that it had created: to more efficiently treat patients admitted to emergency rooms with alcohol poisoning, for example, or to slow the progression of a rare genetic disease called "Fanconi anemia." The real profit to be made from such a drug, though, would presumably have come from the third indication that the company listed on its website: "Alcohol Metabolism Deficiency (ALMD)." This new disease name, which appeared on ALDEA's website and that of a rival pharmaceutical company, is not yet in the International Classification of Diseases.[74] The company did not specify how ALDAs would be used to address this so-called deficiency, but since it is a genetic condition, one assumes that it would require ongoing treatment over the course of an adult lifetime—for as long, as least, as a person afflicted with "ALMD" chose to drink alcohol. ALDEA estimated that "roughly 8% of the world's population (>500 million people)" suffer from this defect, so the potential market was immense.

The entrepreneurs at ALDEA are not the only ones who have been striving to intervene in the kind of alcohol metabolism they consider faulty. At the same time that research journals have been tolling warning bells about how ALDH2*2 elevates one's risk of serious diseases, Silicon Valley startups have highlighted how inconvenient it is, in the work-hard, party-hard culture of young professionals, to have a sudden and visible reaction when drinking alcohol. These new companies draw on the science of alcohol metabolism to offer prophylactic (and, naturally, proprietary) treatments meant to help drinkers digest alcohol more quickly, thereby decreasing the pain of hangovers, the risk of alcohol poisoning, and in some cases, "Asian Flush." Their approaches differ. ZBiotics, the "Pre-Alcohol Probiotic," contains a bacterium genetically engineered to express ALDH that will help break down acetaldehyde faster.[75] H-PROOF, aimed at older people who want to continue to drink like young ones as their bodies start producing less ALDH, claims their tablets will boost ethanol metabolism and protect against liver damage.[76] Another company, the creation of two Singaporean entrepreneurs, takes aim specifically at "Asian Flush." This one, whose moniker DrinkAid makes clear what it is meant to facilitate, incorporates various antioxidants in a tablet to be chewed both before and after drinking. It promises to "reduce Asian flush/glow," "ease discomfort during/after drinking," and "improve productivity the next day."[77]

Some of these companies pay lip service to the idea that drinking less or abstaining from drinking are the most effective ways to avoid the damage alcohol can do to one's body and productivity, but the fact that their success relies on consumers who sicken themselves by drinking makes such statements seem insincere. Because these remedies are sold as supplements, none is subject to FDA approval and the expensive, intensive evaluation process that requires. By comparison, ALDEA's ALDAs faced a longer and more laborious path to approval by the FDA. This may be why, despite raising tens of millions of dollars in funding, the startup seems to have folded.[78] To turn a robust profit, the creators of pharmaceutical remedies will need to convince the medical establishment that alcohol metabolism disorder is a real disease and one whose associated risks make it worthwhile to prescribe a new drug on an ongoing basis.

At first glance, the case of aldehyde dehydrogenase deficiency looks like another example of a phenomenon that many scholars have observed: People who depend on pharmaceutical-company or venture-capital support for their livelihoods have a vested interest in bringing attention to diseases for which they can provide a remedy. In some circumstances this means exaggerating the extent or danger of a particular condition. In others, it may mean turning experiences that are part of a complex human life or dispositions that constitute meaningful elements of self into drug-treatable diseases. It also has the hallmarks of what Robert Aronowitz has called "risky medicine" in that ALMD researchers approach reducing the risk of future diseases as an equivalent priority to treating present symptoms, melding risk with disease.[79]

The present example is even more pernicious than the average example of medicalization, however. For one thing, it is another example of nutritional imperialism, showing how remarkably persistent this approach to diets and bodies remains. Like the example of lactase impersistence, here is a case in which a physiological characteristic of many Chinese bodies—a widespread variant of an enzyme-producing gene—is treated as a defect. Turning the ALDH2*2 allele into alcohol metabolism deficiency creates the impression that many Chinese bodies are sick or weak and need pharmaceutical correction, an impression that serves the interests not only of pharmaceutical companies but of alcohol manufacturers as well.

That raises a further problem: In this case, the proposed treatment may itself exacerbate a public-health problem by facilitating more consumption of alcohol. If the correlation that researchers have noted between carrying the ALDH2*2 allele and developing certain kinds of chronic diseases bears out, then "correcting" alcohol metabolism deficiency could contribute to reducing disease burden in East Asia—at least among people who drink alcohol. But correcting this so-called deficiency is also likely to contribute to an increase in alcohol dependence and attendant health problems by removing the physiological barrier that prevents some people of East Asian descent from drinking alcohol comfortably. Most ALDH2*2-associated debility and disease stems from the *combination* of genes with behavior: If a person with at least one ALDH2*2 allele does not drink alcohol, their risk of alcohol-related cancers is no higher than that of any other non-drinker. The therapies created to date attempt to alleviate the embarrassing, painful, and dangerous immediate reaction that flushers experience when drinking alcohol. Might they not actually encourage more drinking and thereby *increase* the ALDH2*2-associated risk of long-term disease?

Even with the ALDH2*2 obstacle in place, rates of alcohol intake have increased in China across the past few decades. A study published in the *Lancet* in 2019 that analyzed alcohol-exposure data from the World Health Organization found that the global pattern of alcohol use was changing. Alcohol intake per capita had been highest in European countries for as long as such records have been kept, but the data showed that since 1990, alcohol use in Europe had been decreasing, while in China it had been rapidly increasing. The report predicted that by 2030 the East and South Asian regions would eclipse Europe in per-capita alcohol use.[80] China's alcohol-related health problems have increased in tandem; studies from the 2010s have found the rate of alcohol-use disorders in China are approaching the rates in the United States and "other western countries."[81]

What has caused this increase in Chinese alcohol consumption in recent years? More liberal economic policies in China since the rise of Deng Xiaoping in the late 1970s raised incomes and standards of living and thus demand for alcohol as well as other goods. During the Mao years, access to beer in particular had been "a form of urban privilege in a society defined by scarcity," as the historians Jeffrey Pilcher, Wang Yu, and Yuebin Jackson Guo have put it, so once more Chinese people had more money, beer was high on the list of consumer products they desired.[82] But greater purchasing power among PRC citizens is not the only factor at work. In both China and Taiwan, postwar government monopolies on the production and sale of alcohol, which persisted for decades, have ended. In China they ended in the late 1970s as reform-and-opening policies began to take hold, and in

Taiwan only in 2002 with accession to the World Trade Organization (WTO).[83] The agreements to admit China and Taiwan to the WTO in 2001 and 2002, respectively, reduced and eliminated tariffs on imported alcoholic products; the amount of imported alcohol sold increased correspondingly.

Thus, in a sense, today's state of affairs resembles the one that prevailed in the early twentieth century. European, American, and Japanese producers of beer, wine, and liquor are keen to strike it rich in the Chinese market, just as they were 100 years ago. The WTO has smoothed their path to profit by reducing trade barriers, but a genetically determined sensitivity to alcohol could dim their prospects. Thus, this inherited disposition is framed as a disability, one that limits access to professional networking and increases the long-term risk of chronic disease. Compare this to the rhetoric in the past, in which alcohol was presented as modern, invigorating, and health-promoting, and the perceived Chinese preference for opium as consistent with a national failure to thrive. The new narrative, like the old one, implies that deficient drinkers jeopardize national success and prosperity.

Collectively, then, we have returned to something like the early twentieth-century prejudice that framed Chinese temperance as a sign of weakness or inadequacy. The difference is that now scientific research is complicit in the project of reshaping Chinese habits and bodies. We are armed with more precise scientific knowledge, but still limited by a tendency to see variation as deviance and to accept White bodies as the norm.

How to Reframe a Faulty Narrative

Is there a better way to use scientific knowledge about alcohol metabolism, one that doesn't recapitulate 100-year-old prejudices? In the twenty-first century, some scientists have challenged the assumption that carrying an ALDH2*2 allele is a kind of defect. Not all of their approaches are equally useful, however.

A bad way to reframe the narrative is to invert the implied hierarchy and suggest that the ALDH2*2 allele makes one healthier while the more-typical ALDH2*1 allele makes one more prone to sickness. For instance, in 2013 a group of scientists in southwest China tested the genomes of members of the Yi minority group and compared them with those of Han Chinese and found that while nearly 20% of the Chinese subjects had at least one ALDH2*2 allele, none of the Yi subjects did. They speculated that this genetic difference might help explain "the relatively higher risk among the Yi of developing alcohol dependence."[84] In this context, unlike in the English-language research reports described earlier, the scientists presented the Chinese gene pool as superior and the genomes of non-Chinese people as pathogenic. Not only the Yi people, but also "Caucasians, Black Americans, Europeans, and North American Indians" suffered the consequences of lacking the ALDH2*2 variant, according to the Yunnan researchers. The genetic disposition that Stanford laboratories and pharmaceutical-company websites made into a danger here became an asset.

Similarly, a 2018 article published in a cancer-research journal by two scientists in Guangxi Province asks whether ALDH2*2 might be "the reason for low incidence and mortality of ovarian cancer for East Asian women." In a provocative "Final remark," the authors demonstrate that they are just as capable as any early twentieth-century imperialist of extrapolating from one shard of research to explain complex geopolitical phenomena. They write, "East Asia with its ALDH2*2 mutation seems to [have been] the most economically active place in the world" for most of human history, which raises the question "whether ALDH2*2 mutation is helpful to intelligent development."[85]

Observers writing for a less-specialized audience than the Yunnan and Guangxi researchers have felt even freer to speculate about the flushing gene. In the magazine *Health Management* (*Jiankang guanli* 健康管理), a professor with appointments at the prestigious Chinese Academy of Medical Sciences and Peking Union Medical College, Huang Jianshi 黄建始, offered his own unusual explanation of why East Asian and European genomes diverged when it came to alcohol metabolism. Though he is an infectious-disease physician, not a geneticist, his lack of specialized expertise did not prevent him from expounding an authoritative-sounding account according to which flushing was the original, default human reaction to ingesting alcohol. All ancient humans, in this telling, flushed and responded to alcohol as to a poison. This changed, however, as people in Europe and people in East Asia devised different solutions to the universal problem of unsanitary drinking water. East Asians, according to Huang, boiled their drinking water, while Europeans instead turned to fermented beverages such as beer. Since they were using alcoholic beverages for everyday hydration, Europeans had to develop a robust tolerance for alcohol and therefore gradually ceased flushing. Chinese people, who drank alcohol less regularly, never needed to develop the same tolerance and continued to flush as before.[86]

It is a purely speculative argument, untethered to evidence either genetic or historical. By the time this article was published in 2011, scientists had known for at least a decade that ALDH2*2 is the mutation and ALDH2*1 the original pattern, not the reverse as Huang indicates. Still, its appeal is obvious. Like the early lactose-intolerance research discussed in chapter 4, this explanation takes as a norm the physiology of people whose heritage resembles the author's and dismisses as aberrant the physiology of people less like the author. In both cases, this perspective turns out to be wrong: The vast majority of human beings on the planet share a physiology unlike that of the author or scientist. Most human beings are lactose intolerant, and most do not flush when they drink.

While it is useful to resist the idea that processing alcohol differently from the majority is a disorder (resisting that idea is, after all, what this chapter means to encourage), reframing it as a norm or a form of superiority, as these studies do, is not the way forward. Studies and speculations such as the ones just cited simply change the identities of Self and Other in the comparison. Instead of Chinese bodies being prone to pathology, other people's bodies (Yi/White/Native American/Black) are prone to pathology, and instead of Chinese bodies having mutated from some original pattern, other bodies deviated.

One thing these flawed articles get right, however, is this: Reframing this narrative well requires us to remove, or at least explicitly qualify, the taint of pathology that hovers around this genetic constitution. Too often when scientists and those popularizing scientific research discuss long-term health risks associated with ALDH2*2, they fail to emphasize—or sometimes even mention—that these risks emerge in a person who regularly drinks alcohol and not simply from having an ALDH2*2 allele. They normalize the presence of abundant alcohol in a daily diet.

A more promising example of how researchers might resist pathologizing this variation in alcohol metabolism is to explore, as some anthropologists have done, what other health effects the ALDH2*2 allele might have. Observing how quickly the mutation had spread and how thoroughly it had penetrated the gene pool in East Asia, some researchers have wondered whether having it constituted an evolutionary advantage of some kind.

They soon noticed that populations with high proportions of ALDH2*2 alleles also had high rates of hepatitis B. Since drinking alcohol exacerbates the risk of liver failure and liver cancer—a risk already elevated in hepatitis B-infected individuals—researchers speculated that a gene that discouraged alcohol abuse would have been even more beneficial in a population with endemic hepatitis B than in other contexts. They thought that might explain the remarkable success of the ALDH2*2 mutation over a relatively short period.[87] Further, in 2023, a link between ALDH2*2 and resistance to tuberculosis was proposed.[88] Perhaps this mutation that seems at first glance to have been disadvantageous because of the way it increased the risk of certain kinds of diseases was in the bigger picture advantageous because of the protection it offered against diseases more common in the population. In other words, perhaps it served a purpose among East Asian people similar to the one that the mutation causing sickle-cell anemia in some African populations served by protecting against malaria.

Unfortunately, even this sort of research often dresses in race or ethnicity: "Why Can't Chinese Han Drink Alcohol?" asks the title of a journal article on the hepatitis B connection. "What If There's a Secret Benefit to Getting Asian Glow?" is the title of a piece on the tuberculosis hypothesis in the *Atlantic*.[89] This raises the last, indispensable piece of reframing: We must deracialize difference and resist the temptation to read national or racial fitness into what is really just an interesting genetic variation in the human family. Terms like "Asian Flush" and "Asian Glow" homogenize a large and very diverse group of people. They make a genetic disposition characteristic of a minority of that large and diverse group representative of the whole. They also obscure the complex ways in which that genetic disposition manifests. For instance, people who are homozygous ALDH2*2 have much stronger reactions than those who are heterozygous. And they imply that sensitivity to alcohol is unique to people who identify as East Asian when in fact people from other parts of the world and from different backgrounds flush for different genetic reasons.

In the end, it may be the words of a sixteenth-century Chinese scholar that offer the best guide to help us make use of this new scientific knowledge. Writing about ale, Liu Hui 劉繪 (1505–1578) observed that

> those who like it think it brings equanimity; those who don't like it think it is a madness-inducing drug. Those who like it associate it with the holy and wise;

> those who don't like it think of it as evil. Those who like it consider it a shining sea of lustrous pearls; those who don't like it consider it a muddy bed of dirty rocks. Those who like it think it clarifies one's inner spirit like a jade mirror; those who don't, think it offends the nose like a sludge of dregs.

Therefore, Liu concluded, "The good and evil of ale is the result of the person; it doesn't come from the ale itself."[90] It is a simple observation that applies equally well today: People, and bodies, differ in ways that make alcohol suit one person and not the next.

Conclusion

That White Europeans and Americans said obnoxious things about Chinese people in the nineteenth and early twentieth centuries should surprise no one. The hubristic bluster issuing from English-language newspapers and magazines around the turn of the twentieth century is familiar to anybody who has studied imperialism. As Britain, France, Germany, the United States, and other nations expanded their influence around the globe, many involved in the imperial project began to view Asia with contempt. The lopsided distribution of power seemed like confirmation that those in power had some intrinsic gift that those without it lacked. More and more published works depicted Chinese people as passive, meek, effete, and impractical, crafting stereotypes that were a mirror image of the stereotypes they simultaneously built around Western civilization. Both sets of impressions collapsed large, diverse groups of people with complicated histories into one-dimensional caricatures—that is what stereotypes do, after all.

What is more surprising, perhaps, is that these beliefs resonated in nutrition science. Scientists also saw the lopsided distribution of power and wealth and wanted to understand it and, in the case of Chinese scientists, to rectify it. To tackle a problem of such scale and urgency, they naturally turned to what they knew best and applied the tools of their own trade. What was it, they asked, about Western eating patterns that had made Western nations so formidable? And what, by contrast, was missing from the diets of people in China?

Once framed that way, the question prompted any number of hypotheses because global eating patterns had changed dramatically during the nineteenth century, creating marked contrasts between rich industrialized countries and poorer agricultural ones. Milk and beer flowed liberally across the American landscape at a time when industrial dairies and breweries were just beginning to appear in China's treaty ports. Beef was common on British tables when most Chinese eaters could afford meat only a few times a year. Nor had diets in agricultural nations simply remained the same while those in industrialized countries improved. The quality of diets in southern China degraded, for example, as cash cropping spread and more and more people relied for subsistence on imported, steam-milled rice stripped of vitamins.

The scene that the early twentieth century presented to observers, in other words, was not one in which people in China had access to the same range of foodstuffs as people in rich countries but perversely chose to eat less nutritious things. The main problem was poverty, not poor choices. And traditional food beliefs and practices had little to do with comparative states of health, size, or longevity.

Many nutrition scientists were aware of this, as their work occasionally reveals. Hsien Wu (Wu Xian 吳憲) once wrote that while he thought Chinese people ate too little animal protein, there was no hope of improving their nutritional status until more people could get more calories in their stomachs, period.[1] But the scientists' tools were better suited to questions about food choice than to questions about justice. So they beavered away, deconstructing typically Chinese foodstuffs to determine how nutrient-rich or nutrient-poor they were, raising generation after generation of rats on controlled rations, and measuring nitrogen levels in the urine of college students.

This outpouring of effort and brilliance produced many admirable results, not least the foundation of a new scientific discipline in China. But because the question underlying the research was so often "What is wrong with China?", that research also legitimized the stereotypes that unabashed defenders of imperialism had propagated. Chinese diets were vitamin- and protein-deficient; no wonder Chinese people were so weak and passive. Chinese people could not drink milk; no wonder they were so small. In most cases it had never been the scientists' intention to prop up racial prejudice. In fact, for many of them dietary explanations appealed precisely because they emphasized the possibility of growth and change in a world too ready to ascribe fixed characteristics to people of different nations or races. But in practice their work ended up pathologizing Chinese bodies all the same.

They also misrepresented China's dietary history as well as the histories of European nations and the United States, reading contrasts that were actually modern back into the mists of antiquity. Compared to Europeans and their descendants, Chinese people did not appreciate milk, eat wheat or meat, or get drunk: Such claims might have been somewhat plausible in particular places at the peculiar moment that was the early twentieth century, but even a cursory reading of premodern sources does not support them. One cannot fault scientists for not being historians, of course, except that these scientists sometimes made explicit claims about history—that Buddhism had stripped Chinese diets of animal protein, for example, or that dairy had never been a part of Chinese culinary or dietetic culture.

Why did nutritional imperialism not simply fade away as formal empires (mostly) did starting in the mid-twentieth century? The examples in this book suggest a couple of explanations. First of all, European nations and the United States continued to wield disproportionate power and influence globally even as decolonization proceeded. They dominated the United Nations and its constituent organizations, including the Food and Agriculture Organization (FAO). Think of the United States, the United Kingdom, the Netherlands, and Australia successfully arguing in the 1940s that they deserved as much say in the FAO's Rice Study Group as the rice-producing nations of Asia.

Then there are the financial motivations. In the second half of the twentieth century Taiwan and—eventually—the PRC became attractive targets as new markets for European and American agricultural surpluses. The United States arranged to sell its farmers' excess wheat and dairy to Taiwan at a discount, an arrangement that benefited both American farmers and the Taiwanese government. Both groups then had a material stake in promoting the idea that what Americans produced in abundance were the most nutritious kinds of foods. As the market for these specific foodstuffs grew, Taiwan also developed its productive capacity, adding its own farmers and entrepreneurs to the list of people invested in stoking demand for the sorts of foods Americans had earlier been unloading.

In the People's Republic the dynamic was different during the Mao years, but after China adopted more liberal economic policies in the late 1970s, a similar set of motivations came into play. Chinese consumers started to become a market—a potentially huge one—for wheat, milk, meat, and alcohol to the benefit of both foreign and Chinese companies.

China's and Taiwan's accession to the World Trade Organization at the dawn of the twenty-first century removed many of the remaining barriers to the importation of foreign food and drink, such as Taiwan's government monopoly on alcohol. And then, of course, there are the financial incentives for companies that produce remedies or prophylaxis—low-lactose dairy products or pills to cure "alcohol metabolism disorder"—that can enable everyone, regardless of their bodily differences, to consume large amounts of the approved things. In short, today a lot of people have a lot invested in the perception that Chinese diets and metabolism require reform.

Finally, as the examples of lactose intolerance and Asian Flush show, the ways in which medical research and practice have changed in recent years can also shore up nutritional imperialism. The rise of what the historians Robert Aronowitz and Jeremy Greene have called "risky medicine" and "prescribing by numbers," respectively, has accustomed us to the idea that we might be ill even if we don't feel like we are, even if we're suffering no obvious symptom or dysfunction. This is the twenty-first-century version of the New Nutrition claim in the 1920s that subclinical nutritional deficits were a greater threat to a nation than symptomatic vitamin deficiency disease, and that adequate diet was not enough—optimum diet was what a society ought to strive for. In the throes of such a belief, it is easy to accept that a genetic variation in metabolism is a health impediment that requires intervention.

More alarming is how genomic medicine and "precision" or "personalized" nutrition—that is, nutritional advice purportedly derived from an individual's characteristics rather than general population-level data—tend to recapitulate the pitfalls of early twentieth-century racial science. The promise of precision nutrition is that by amassing more genetic data on groups of people underrepresented in past samples, nutritionists will be able to provide advice more precisely suited to, say, clients of Chinese descent than they might have been able to do when relying on older datasets. Like metabolism experiments a hundred years ago, however, this approach prioritizes difference and casts in racial terms any differences that researchers discover in the group they are studying. Racial metabolism researchers in chapter 1 expected a "Chinese normal" different from the "standard man" based on European and American bodies. The difference that they thought they'd found—that "Oriental" bodies processed food more slowly than "Occidental" ones—was used to legitimize proposals to ration food differently based on race. While policymakers will probably not use precision nutrition in ways as egregious as that, this kind of research is nonetheless likelier to harm than to help the cause of racial equality.

That is because precision nutrition focuses on reforming the individual at the expense of addressing the social and economic determinants of health. To some degree early twentieth-century Chinese nutrition scientists did this, too. Their various diagnoses of what Chinese eaters were doing wrong—they avoided meat, they hadn't prioritized milk, they ate rice instead of wheat—seem myopic given the environment of widespread poverty and continuous warfare that surrounded them. Today, likewise, if the circumstances of a person's life prevent them from accessing varied and healthy food, they will not be well nourished, whatever their genetic disposition. Investing time, money, and brains in more precisely mapping the genetic foundations of metabolic difference not only distracts from the more significant causes of poor nutrition but also reinforces a message that the problem lies in the individual—in their genome, in their lifestyle, in their dietary choices. The collective problem, by extension, comes to be seen as inherent in the larger group with which the individual is identified, whether that is Chinese, Asian, Black, or any other label. Today's genomic nutrition is likely to reify some of the race-based concepts of early twentieth-century science and is no more likely to improve health status in the aggregate.

"But wait," one might object. "Chinese people's health status *has* improved, a lot, since the early twentieth century. Doesn't that prove that that early nutrition science—even if it unfortunately reinforced some stereotypes—succeeded?" There is no doubt that people in China and Taiwan are now healthier in the aggregate than they had been in the early and mid-twentieth century. Average height has shot up, for example, and life expectancy has risen. This may appear to affirm the conviction that fixing Chinese diets' specific deficiencies has worked. The argument would be more convincing, however, if all other things had remained constant—if the only thing that had changed in the lives of Taiwanese people had been that they started drinking milk, for example, or if a Chinese teenager in the twenty-first century lived just the same way as their grandmother had in 1965, apart from eating more wheat than rice. But Chinese lives have changed wholesale over the past half-century. On average people are richer and safer, and the food they eat is cleaner, more diverse, and more abundant. Public-health infrastructure has improved enormously. No one product, be it wheat, meat, or milk (and certainly not alcohol) can take credit for the result that they are also healthier.

To summarize, then, the elements that created and have sustained nutritional imperialism in the Chinese case are racist and Orientalist attitudes common in the early twentieth century; Chinese scientists' desire to contribute to their nation's revival and thriving; insufficient awareness of historical changes in diet; the continuing influence of American and European states in the postwar global order; the economic interests of farmers, companies, and governments; and changes in the orientation of nutrition and disease research over the past few decades.

The particular manifestations of nutritional imperialism in China are unique, but the underlying dynamics are not, so I suspect that those versed in the modern history of other parts of the world—India, Japan, southeast Asia, former African colonies, and Latin America, for example—will be able to see its contours there, too. Comparing the Chinese case with others will help us understand the phenomenon better. Japan, for example, had a distinctive status in the age of European imperialism in that it was the only East Asian imperial power. How did that influence the form of nutritional imperialism there? In the case of India, did nutritional imperialism develop differently there because India was a formal colony, as Shang-Jen Li has hinted?[2]

Moreover, there is no reason to believe that nutritional imperialism could originate only in the European-dominated imperial order of the nineteenth and early twentieth centuries. As the philosopher Ian Hacking has observed, the "European empires and their American continuation" deserve special scrutiny because they shaped institutions and ideas that we are only slowly and with great difficulty reshaping today.[3] But obviously it is not the case that only Europeans and people of European descent can be imperialists. Consider, for example, recent reports that the Chinese government has been attempting to teach Uyghur housewives to make and appreciate Chinese cuisine and to replace Uyghur-style breakfasts with Chinese-style ones.[4] Officials' insistence that this amounts to cultivating "scientific and healthy" habits among people who have been eating badly resonates with early-twentieth-century reformers' determination to get Chinese eaters to reject rice and turn to wheat. Can genetic research showing that Uyghurs are congee intolerant be far behind?

My hope, in short, is that by analyzing one example of a pattern, this book has invited more research that will build on, modify, and refine the concept. I have found the idea of nutritional imperialism a useful lens through which to view modern dietetic knowledge in China, and perhaps it can serve as a tool in other scholars' kits as well.

As for the specific incarnation of nutritional imperialism whose history this book has chronicled, there are signs that it may be changing. Organized resistance has emerged. Taiwanese researchers, for example, have lately been advancing the idea of a Taiwanese eating approach (cleverly abbreviated TEA), a "mortality-lowering total wellbeing dietary pattern consistent with the current knowledge" that centers distinctively East Asian elements such as soy products, tea, and boiling or steaming rather than frying or baking. This is meant to both emulate and compete with the branding success of the Mediterranean diet, a concept based on southern European eating habits said to be particularly health-promoting. Among other things, TEA's proponents recommend revising official Taiwanese dietary guidelines to emphasize eating more plant-based than animal-based protein and to walk back the recommendation that everyone drink one to two glasses of milk daily, which "seems unrealistic for the Taiwanese ethnicity."[5]

Similarly, in mainland China discussion of the "Eastern diet model" (*Dongfang shanshi moshi* 东方膳食模式) has heated up since the 2022 version of *Chinese Residents' Dietary Guidelines* (*Zhongguo jumin shanshi zhinan* 中国居民膳食指南) promoted the concept of a distinctively Chinese diet superior to those originating elsewhere. One article describing the Eastern diet began, "In the past it seems that the diet models we've heard about have all been 'imports,' for example 'the Mediterranean diet,' 'DASH [Dietary Approaches to Stop Hypertension] diet,' Japanese diet, and so forth, as if healthy food and drink were all in other countries, but in reality it is not like this. Actually, since ancient times the Chinese people have had a strong awareness of healthy eating, and every place in the country has a food tradition that benefits health." The problem, the author continued, was that unhealthy habits had crept into mainstream diets as the economy developed, so "lots of healthy folk diet traditions were forgotten."[6] Unlike TEA, the Eastern diet does not center characteristically Chinese foodstuffs such as soy products or green tea, and it does not recommend reducing dairy requirements or eating more plant-derived than animal-derived protein.

Suggestions to rebalance the sources of protein have cropped up in mainland China as well, however. In 2012 the chairman of a corporation that makes soy products proposed to China's National People's Congress that in order to promote health and reduce chronic-disease risk, China ought to adopt a "Dual Protein with Chinese Characteristics" program, encouraging consumers to "take vegetable protein as primary, and animal protein as supplementary."[7] This inverts the 1920s admiration for Western diets in which animal protein constituted a much greater proportion of overall dietary protein than it did in China at the time. Since then, the "Dual Protein with Chinese Characteristics" idea has made its way into the "Citizens' Nutrition Plan 2017–2030" put together by the State Council's General Office, though the officials formulating the plan implicitly rejected the soy baron's suggestion to prioritize vegetable protein. In the official document "dual protein" means simply "taking good-quality animal and vegetable protein as the main nutritional material"—a formulation that suggests that there is still some ambivalence about elevating vegetable over animal protein after so many decades of seeing vegetable protein as inferior and a cause of national weakness.[8] Nevertheless, officials have recently made public speeches alluding to a need to rectify "imbalanced consumption" of protein. It seems clear that a change in dietary recommendations, branded as intrinsically Chinese, is underway.[9]

Chinese and Taiwanese voices are not the only ones that have been valorizing Chinese eating patterns in recent decades. Foreigners have, too. The most well-known example is T. Colin Campbell, a Cornell University professor who in 2005 published a book with his son called *The China Study*. The subtitle—*The Most Comprehensive Study of Nutrition Ever Conducted and the Startling Implications for Diet, Weight Loss and Long-Term Health*—gives a sense of the book's tenor. Based partly on Chinese cancer atlases and a large-scale study conducted in Chinese counties in the mid-1970s, then followed up in the mid-1980s, the book argues that a diet mostly free of meat, milk, and eggs like the ones eaten in the counties studied protects against chronic diseases such as heart disease, diabetes, and cancer.[10] It quickly became one of the best-selling books on nutrition ever published in the United States. Other signs that Americans' valuation of "Chinese diet," however defined, has swung from disparagement to admiration is the appearance of English-language books explaining ideas from classical Chinese dietetics, such as Bob Flaws's *The Tao of Healthy Eating*, Joerg Kastner's *Chinese Nutrition Therapy*, or *Functional Foods of the East*, published by CRC Press in 2010.[11] There have even been cookbooks based on Chinese dietetic theory, such as *The Chinese Medicine Cookbook: Nourishing Recipes to Heal and Thrive*, published in 2019.[12]

What accounts for this recent reversal of historically negative perceptions of Chinese diets? One answer is political. The global disposition of wealth and authority has changed. China and Taiwan have both grown richer and more politically influential than they were in the early twentieth century. With that growth has come—particularly on the part of the PRC—a greater readiness to reject European and American models as norms. Moreover, the growing representation of Chinese and Taiwanese immigrants and their descendants in North America and European countries since the Second World War has brought more diverse perspectives and greater awareness of Chinese ideas and practices.

A second, equally important answer is epidemiological. In China as elsewhere, as the burden of disease has shifted from acute infectious to chronic noncommunicable diseases, nutrition experts have begun to question recommendations to eat more meat and dairy. As rates of heart disease and diabetes rose in North America and Europe in the mid-twentieth century and since the 1990s as public health experts have made managing obesity a top priority, earlier boasts that so-called Western diets empower nations have become laments that Western diets enfeeble them. At the same time, international interest in diets considered traditional began to increase. Therefore, the "Chinese diet," which had previously been considered a cause of weakness and corruption in Chinese bodies and spirits, became instead a source of health and longevity in contrast to the "Western diet," said to induce chronic disease and early death.

This form of pushback does not show that we have escaped the confines of nutritional imperialism. It simply confirms how profoundly political, economic, and epidemiological contexts influence our understanding of what is good (or not good) to eat. Both the older assessment and the new one draw on simplistic and inaccurate notions that there is such a thing as a uniform Chinese or Western diet. Both tend to homogenize not only diets but also people and to make empowering the nation a central goal of dietary research. In the new assessment, an imagined China remains the foil of an imagined West, but it is understood as superior rather than inferior. Dietetic research questions inspired by Chinese nationalism are no more likely to be useful than those inspired by European and American imperialism.

Considering this, we might do well to also consider an admonition attributed to the historical Buddha, Siddhartha Gautama—namely, some questions do not lead to enlightenment. A well-known story from the Pali canon describes an exchange between the Buddha and his disciple Mālunkyaputta in which the disciple asked his teacher for answers to a number of metaphysical questions. The disciple had made up his mind to leave the religious life if his teacher could not or would not answer these questions. "Is the world eternal, or not?" he asked. "Is the world finite, or not? Is the life principle the same as the body? Does the Buddha exist after death?" The Buddha replied that by insisting on answers to these questions before persisting in the religious life, Mālunkyaputta was like a man shot with a poisoned arrow who refuses help until he's gotten answers to his queries about who shot the arrow (What color is his skin? How tall is he? Where does he live?), what kind of bow shot it (longbow? crossbow? reed, hemp, or sinew bowstring?), or what kind of feathers were fitted to the shaft (vulture? crow? hawk?). The wounded man will die without getting answers to his questions. Just so, the Buddha told his student, by fixating on these metaphysical questions, you are ignoring the more urgent priority. It is hard enough to reach nirvana as it is; distracting your mind and efforts with questions irrelevant to that purpose will prevent you from achieving it.[13]

In the modern period, many of the questions that scientists have asked about Chinese eaters—Do they process food more slowly than others? Has eating rice or failing to eat meat made them weak? Are they hampered by genetic defects that stymie their ability to digest milk or alcohol?—similarly miss the point. Pursuing questions about the defects of a particular culture's foodways or about racial factors in digestion or metabolism has rarely if ever promoted the main goal of dietetic knowledge—namely, robust human health. Foodways differ intriguingly, but comparing the nutritional value of diets in one culture with those in another compels us to essentialize both cultures and diets. And while different groups apparently have some genetic variations that affect food consumption, such differences are for the most part interesting trivia rather than discoveries with important public-health implications.

Like the man in the Buddha's parable, we have been shot by a poisoned arrow. We live in a world where nutritious food is inequitably distributed, a world where some people suffer from undernutrition because they can't obtain enough calories, while others lack access to much other than highly processed carbohydrates loaded with salt, fat, and sugar. And like the earnest disciple, we fixate on questions incidental to our predicament. While we devote effort and resources to investigating genetic metabolic differences or a traditional diet's nutritional value, the poison continues to spread.

Thus, whenever scientific or political authorities introduce a new dietary recommendation or new digestive "deficiency," we ought to be on alert. If the new concept reinforces stereotypes about a particular racial or ethnic group and requires that they change their foodways or their bodies, we ought to raise an eyebrow. If there's money involved—if changing foodways or bodies would require importing food, developing and selling pharmaceutical products, or lavishly funding new research—that is all the more cause for hesitation. Having recognized the unsavory origins of some seemingly neutral scientific concepts, we can call out nutritional imperialism when we encounter it. Perhaps, over time, we can find our way to dietetic knowledge that recognizes human diversity without pathologizing it.

Suggested Further Reading

Changes in East Asian Cuisine Over Time

Anderson, E. N. *Food and Environment in Early and Medieval China* (Philadelphia: University of Pennsylvania Press, 2014).

Anderson, E. N. *The Food of China* (New Haven, CT: Yale University Press, 1988).

Barstow, Geoffrey. *Food of Sinful Demons: Meat, Vegetarianism, and the Limits of Buddhism in Tibet* (New York: Columbia University Press, 2018).

Chang, K. C., ed., *Food in Chinese Culture: Anthropological and Historical Perspectives* (New Haven, CT: Yale University Press, 1977).

Chen, Yong. *Chop Suey, USA: The Story of Chinese Food in America* (New York: Columbia University Press, 2014).

Coe, Andrew. *Chop Suey: A Cultural History of Chinese Food in the United States* (Oxford: Oxford University Press, 2009).

Dubois, Thomas David. *China in Seven Banquets: A Flavourful History* (London: Reaktion Press, 2024).

Dunlop, Fuchsia. *Invitation to a Banquet: The Story of Chinese Food* (New York: Norton, 2024).

Dunlop, Fuchsia. "The Strange Tale of General Tso's Chicken," in *Authenticity in the Kitchen: Proceedings of the Oxford Symposium on Food and Cookery*, edited by Richard Hosking, 165–177 (Devon, UK: Prospect Books, 2006).

Fu, Jia-Chen. *The Other Milk: Reinventing Soy in Republican China* (Seattle: University of Washington Press, 2018).

Höllman, Thomas O. *The Land of the Five Flavors: A Cultural History of Chinese Cuisine* (New York: Columbia University Press, 2010).

King, Michelle T. *Chop Fry Watch Learn: Fu Pei-Mei and the Making of Modern Chinese Food* (New York: Norton, 2024).

King, Michelle T. "What Is 'Chinese' Food? Historicizing the Concept of Culinary Regionalism." *Global Food History* 6, no. 2 (2020): 89–109.

Lee, Seung-joon. *Gourmets in the Land of Famine: The Culture and Politics of Rice in Modern Canton* (Stanford, CA: Stanford University Press, 2011).

Roberts, J. A. G. *China to Chinatown: Chinese Food in the West* (London: Reaktion Books, 2002).

Sabban, Françoise. "Un savoir-faire oublié: Le travail du lait en Chine ancienne," *Zinbun* no. 21 (1986): 31–65.

Sabban, Françoise. "The Taste for Milk in Modern China (1865–1937)," in *Food Consumption in Global Perspective*, edited by Anne Murcott and Jakob A. Klein, 182–209 (Basingstoke: Palgrave Macmillan, 2014).

Sabban, Françoise. "La viande en Chine: Imaginaire et usages culinaires," *Anthropozoologica* no. 18 (1993): 79–90.

Sterckx, Roel. *Food, Sacrifice, and Sagehood in Early China* (Cambridge: Cambridge University Press, 2011).

Sterckx, Roel, ed. *Of Tripod and Palate: Food, Politics, and Religion in Traditional China* (New York: Palgrave Macmillan, 2005).

Swislocki, Mark. *Culinary Nostalgia: Regional Food Culture and the Urban Experience in Shanghai* (Stanford, CA: Stanford University Press, 2009).

Wang, Q. Edward. *Chopsticks: A Cultural and Culinary History* (Cambridge: Cambridge University Press, 2015).

Wu, David Y. H., and Sidney C. H. Cheung. *The Globalization of Chinese Food* (Honolulu: University of Hawai'i Press, 2002).

Wu, David Y. H., and Chee-beng Tan. *Changing Chinese Foodways in Asia* (Hong Kong: Chinese University Press, 2001).

Famines in Asia at the End of World War II

Aldous, Christopher. "Contesting Famine: Hunger and Nutrition in Occupied Japan, 1945–1952," *Journal of American-East Asian Relations* 17, no. 3 (2010): 230–256.

Gunn, Geoffrey. *Rice Wars in Colonial Vietnam: The Great Famine and the Viet Minh Road to Power* (Lanham, MD: Rowman & Littlefield, 2014).

Huff, G. "Causes and Consequences of the Great Vietnam Famine, 1944–5," *Economic History Review* 72, no. 1 (2019): 286–316.

Ó Gráda, Cormac. "The Famines of WWII," *VoxEU*, September 2, 2019, https://cepr.org/voxeu/columns/famines-wwii.

van der Eng, Pierre. "Food Supply in Java during War and Decolonisation, 1940–1950," Munich Personal RePEc Archive (MPRA) Paper 8852, Second Edition (2008). Internet Archive URL: https://web.archive.org/web/20260115124045/https://mpra.ub.uni-muenchen.de/8852/.

Food and Health in East Asia

Bay, Alexander R. *Beriberi in Modern Japan: The Making of a National Disease* (Rochester, NY: University of Rochester Press, 2012).

Engelhardt, Ute. "Dietetics in Tang China and the First Extant Works of Materia Dietetica," in *Innovation in Chinese Medicine*, edited by Elisabeth Hsu, 173–191 (Cambridge: Cambridge University Press, 2001).

Leung, Angela Ki Che, and Melissa L. Caldwell, eds., *Moral Foods: The Construction of Nutrition and Health in Modern Asia* (Honolulu: University of Hawai'i Press, 2019).

Lo, Vivienne. "Pleasure, Prohibition and Pain: Food and Medicine in Traditional China," in *Of Tripod and Palate: Food, Politics, and Religion in Traditional China*, edited by Roel Sterckx, 163–185 (New York: Palgrave Macmillan, 2005).

Sabban, Françoise. "Transition nutritionelle et histoire de la consummation laitière en Chine," *Cholé-Doc* no. 120 (2010): 1–6.

Smith, Hilary A. "Food, Health, and Nutrition in Chinese History," *History Compass* 20, no. 1 (January 2022): 1-12.

Swislocki, Mark. "Nutritional Governmentality: Food and the Politics of Health in Late Imperial and Republican China," *Radical History Review* no. 110 (2011): 9–35.

Torck, Mathieu. *Avoiding the Dire Straits: An Inquiry into Food Provision and Scurvy in Maritime and Military History of China and Wider East Asia* (Wiesbaden, DE: Harrassowitz Verlag, 2009).

Food, Health, and Disease—General

Cantor, David, Christian Bonah, and Matthias Dörries, eds., *Meat, Medicine, and Human Health in the Twentieth Century* (London: Pickering and Chatto, 2010).

Carpenter, Kenneth J. *Beriberi, White Rice, and Vitamin B: A Disease, a Cause, a Cure* (Berkeley: University of California Press, 2000).

Carpenter, Kenneth J. *The History of Scurvy and Vitamin C* (Cambridge: Cambridge University Press, 1986).

Carpenter, Kenneth J. *Protein and Energy: A Study of Changing Ideas in Nutrition* (Cambridge: Cambridge University Press, 1994).

Etheridge, Elizabeth W. *The Butterfly Caste: A Social History of Pellagra in the South* (Westport, CT: Greenwood, 1972).

Hatch, Anthony Ryan, Sonya Sternlieb, and Julia Gordon, "Sugar Ecologies: Their Metabolic and Racial Ecologies," *Food, Culture & Society* 22, no. 5 (2019), 595–607.

Moran-Thomas, Amy. *Traveling with Sugar: Chronicles of a Global Epidemic* (Berkeley: University of California Press, 2019).

Roe, Daphne A. *A Plague of Corn: The Social History of Pellagra* (Ithaca, NY: Cornell University Press, 1973).

Semba, Richard D. *The Vitamin A Story: Lifting the Shadow of Death* (Basel, CH: Karger, 2012).

Warren, Christian. *Starved for Light: The Long Shadow of Rickets and Vitamin D Deficiency* (Chicago: University of Chicago Press, 2024).

Imperialism, Food and Health in Asia

Arnold, David. *Colonizing the Body: State Medicine and Epidemic Disease in Nineteenth-Century India* (Berkeley: University of California Press, 1993).

Au, Sokhieng. *Mixed Medicines: Health and Culture in French Colonial Cambodia* (Chicago: University of Chicago Press, 2011).

Collingham, E. M. *Imperial Bodies: The Physical Experience of the Raj c. 1800–1947* (Cambridge: Polity Press, 2001).

Jones, Margaret. "Tuberculosis, Housing and the Colonial State: Hong Kong, 1900–1950," *Modern Asian Studies* 37, no. 3 (2003): 653–682.

Leong-Salobir, Cecilia. *Food Culture in Colonial Asia: A Taste of Empire* (New York: Routledge, 2011).

Liu, Michael Shiyung. *Prescribing Colonization: The Role of Medical Practices and Policies in Japan-Ruled Taiwan, 1895–1945* (Ann Arbor, MI: Association for Asian Studies, 2009).

Monnais, Laurence. *The Colonial Life of Pharmaceuticals: Medicines and Modernity in Vietnam* (New York: Cambridge University Press, 2019).

Peckham, Robert. "Hong Kong Junk: Plague and the Economy of Chinese Things," *Bulletin of the History of Medicine* 90, no. 1 (Spring 2016), 32–60.

Peters, Erica J. *Appetites and Aspirations in Vietnam: Food and Drink in the Long Nineteenth Century* (Lanham, MD: AltaMira Press, 2012).

Nutritional Paradigms, Europe and the United States

Albala, Ken. *Eating Right in the Renaissance* (Berkeley: University of California Press, 2002).

Apple, Rima D. *Vitamania: Vitamins in American Culture* (New Brunswick, NJ: Rutgers University Press, 1996).

Biltekoff, Charlotte. *Eating Right in America: The Cultural Politics of Food and Health* (Durham, NC: Duke University Press, 2013).

Ehrenberger, Kristen Ann. *The Politics of the Table: Nutrition and the Body in Modern Germany* (Baltimore, MD: Johns Hopkins University Press, 2026).

Kamminga, Harmke and Andrew Cunningham, eds. *The Science and Culture of Nutrition, 1840–1940* (Atlanta, GA: Rodopi, 1995).

Laas, Molly S. "Nutrition as a Social Question: 1835–1905," PhD dissertation, University of Wisconsin-Madison, 2017.

Levenstein, Harvey. *Revolution at the Table: The Transformation of the American Diet* (Oxford: Oxford University Press, 1988).

Mudry, Jessica J. *Measured Meals: Nutrition in America* (Albany: State University of New York Press, 2009).

Ruis, Andrew R. *Eating to Learn, Learning to Eat: The Origins of School Lunch in the United States* (New Brunswick, NJ: Rutgers University Press, 2017).

Shapin, Steven. *Eating and Being: A History of Ideas about Our Food and Ourselves* (Chicago: University of Chicago Press, 2024).

Shapin, Steven. "Expertise, Common Sense, and the Atkins Diet," in *Public Science in Liberal Democracy*, edited by Jene M. Porter and Peter W. B. Phillips, 174–193 (Toronto: University of Toronto Press, 2007).

Shapin, Steven. "Trusting George Cheyne: Scientific Expertise, Common Sense, and Moral Authority in Early Eighteenth-Century Dietetic Medicine," *Bulletin of the History of Medicine* 77, no. 2 (2003): 263–297.

Shapin, Steven. "Was Luigi Cornaro a Dietary Expert?" *Journal of the History of Medicine and Allied Sciences* 73, no. 2 (April 2018): 135–149.

Shapin, Steven. "'You Are What You Eat': Historical Changes in Ideas about Food and Identity," *Historical Research* 87, no. 237 (August 2014): 377–392.

Spary, E. C. *Eating the Enlightenment: Food and the Sciences in Paris, 1670–1760* (Chicago: University of Chicago Press, 2012).

Tracy, Sarah W. "A Global Journey—Ancel Keys, the FAO, and the Rise of Transnational Heart Disease Epidemiology, 1949–1958," *International History Review* 41, no. 2 (2019): 372–390.

Treitel, Corinna. *Eating Nature in Modern Germany: Food, Agriculture and the Environment, c. 1870 to 2000* (Cambridge: Cambridge University Press, 2017).

Treitel, Corinna. "Nutritional Modernity: The German Case," *Osiris* 35, no. 1 (2020): 183–203.

Notes

Introduction

1. In this book, I capitalize racial and civilizational designations such as "White," "Oriental," and "Black" to emphasize that all are human constructions. Today some scholars and news organizations choose to capitalize some but not others ("Black," for example, but not "white"), but I follow Nell Irvin Painter in capitalizing all in order to avoid implying that some racial identities are natural or normal while others are constructed. Nell Irvin Painter, "Why 'White' Should Be Capitalized Too," *Washington Post*, July 22, 2020). Internet Archive URL: https://web.archive.org/web/20200724040621/https://www.washingtonpost.com/opinions/2020/07/22/why-white-should-be-capitalized/.

2. G. M. Beard, "The Diet of Brain-Workers," *Hours at Home: A Popular Monthly of Instruction and Recreation* 9, no. 5 (September 1869): 423.

3. Beard, "The Diet of Brain-Workers," 423.

4. Hsien Wu (Wu Xian 吳憲), "The Chinese Diet in the Light of Modern Knowledge of Nutrition," *Chinese Social and Political Science Review* 11, no. 1 (January 1927): 74.

5. American Federation of Labor, "Some Reasons for Chinese Exclusion. Meat vs. Rice. American Manhood against Asiatic Coolieism. Which Shall Survive?" (Washington, DC: Government Printing Office, 1902).

6. Shang-Jen Li, "Eating Well in China: British Medical Men on Diet and Personal Hygiene in Nineteenth-Century Chinese Treaty Ports," in *Health and Hygiene in Chinese East Asia: Policies and Publics in the Long Twentieth Century*, edited by Angela Ki Che Leung and Charlotte Furth, 109–131 (Durham, NC: Duke University Press, 2010).

7. Laura Kurtzman, "Trump's 'Chinese Virus' Tweet Linked to Rise of Anti-Asian Hashtags on Twitter," University of California San Francisco News, March 18, 2021. Internet Archive URL: https://web.archive.org/web/20210319063827/https://www.ucsf.edu/news/2021/03/420081/trumps-chinese-virus-tweet-linked-rise-anti-asian-hashtags-twitter.

8. Edward G. A. Iglesia, "Aldehyde Dehydrogenase Deficiency," in *Encyclopedia of Food Allergy*, vol. 1, edited by Scott H. Sicherer (Amsterdam: Elsevier, 2024), 732.

9. Projit Bihari Mukharji, *Brown Skins, White Coats: Race Science in India, 1920–1966* (Chicago: University of Chicago Press, 2022).

10. Lillian M. Li, *Fighting Famine in North China: State, Market, and Environmental Decline, 1690s–1990s* (Stanford, CA: Stanford University Press, 2007); Kathryn Edgerton-Tarpley, *Tears from Iron: Cultural Responses to Famine in Nineteenth-Century China* (Berkeley: University of California Press, 2008); Mike Davis, *Late Victorian Holocausts: El Niño Famines and the Making of the Third World* (London: Verso, 2001); Kenneth Pomeranz, "Political Economy and Ecology on the Eve of Industrialization: Europe, China, and the Global Conjuncture," *American Historical Review* 107, no. 2 (2002): 425–446.

11. Ken Albala, "Japanese Food in the Early Modern European Imagination," in *Devouring Japan: Global Perspectives on Japanese Culinary Identity*, edited by Nancy Stalker, 35–47 (Oxford: Oxford University Press, 2018).

12. Li, "Eating Well in China"; David Arnold, "The 'Discovery' of Malnutrition and Diet in Colonial India," *Indian Economic and Social History Review* 31, no. 1 (1994): 1–26.

13. Wang Gong 王公 and Yang Jian 杨舰, "Kang zhan yingyang baozhang tixi de jianli yu Zhongguo yingyangxue de jianzhihua 抗战营养保障体系的建立与中国营养学的建制化" [On the establishment of wartime nutrition guarantee system and the institutionalization of modern nutrition science in China during World War II], *Ziran bianzhengfa tongxun* 自然辩证法通讯 [Journal of the dialectics of nature] 41, no. 8 (2019): 62–70.

14. *Henry Lester Institute of Medical Research Annual Report* (Shanghai: Henry Lester Institute, 1935); Zheng Ji 郑集 (Libin T. Cheng), *Zhongguo zao qi shengwu huaxue fazhan shi* 中国早期生物化学发展史 1917–1949 [The history of the early development of biochemistry in China 1917–1949] (Nanjing: Nanjing University Press, 1989).

15. Wang Feng, *China's Age of Abundance: Origins, Ascendance, and Aftermath* (Cambridge: Cambridge University Press, 2024), 60–81; Elizabeth J. Leppman, *Changing Rice Bowl: Economic Development and Diet in China* (Hong Kong: Hong Kong University Press, 2005).

16. See David Arnold, "Introduction: Disease, Medicine and Empire," in his *Imperial Medicine and Indigenous Societies* (Manchester: Manchester University Press, 1988).

17. Alfred W. Crosby, *Ecological Imperialism: The Biological Expansion of Europe 900–1900* (Cambridge: Cambridge University Press, 1986); William H. McNeill, *Plagues and Peoples* (New York: Anchor Books, 1976).

18. See Daniel R. Headrick, *The Tools of Empire: Technology and European Imperialism in the Nineteenth Century* (New York: Oxford University Press, 1981); Roy MacLeod and Milton Lewis, eds., *Disease, Medicine, and Empire: Perspectives on Western Medicine and the Experience of European Expansion* (London: Routledge, 1988); Arnold, *Imperial Medicine and Indigenous Societies*.

19. Amartya Sen, *Poverty and Famines: An Essay on Entitlement and Deprivation* (Oxford: Clarendon Press, 1981); Diana Wylie, *Starving on a Full Stomach: Hunger and the Triumph of Cultural Racism in Modern South Africa* (Charlottesville: University Press of Virginia, 2001); Davis, *Late Victorian Holocausts*.

20. For example, Mary P. Sutphen and Bridie J. Andrews, eds., *Medicine and Colonial Identity* (New York: Taylor and Francis, 2003); Sokhieng Au, *Mixed Medicines: Health and Culture in French Colonial Cambodia* (Chicago: University of Chicago Press, 2011); Laurence Monnais, *The Colonial Life of Pharmaceuticals: Medicines and Modernity in Vietnam* (New York: Cambridge University Press, 2019).

21. Rachel Laudan, *Cuisine and Empire: Cooking in World History* (Berkeley: University of California Press, 2013). See also Rebecca Earle, *The Body of the Conquistador: Food, Race and the Colonial Experience in Spanish America, 1492–1700* (Cambridge: Cambridge University Press, 2012).

22. Michael Worboys, "The Discovery of Colonial Malnutrition between the Wars," in *Imperial Medicine and Indigenous Societies*, edited by David Arnold, 208–225 (Manchester: Manchester University Press, 1988).

23. Wylie, *Starving on a Full Stomach*; Paul Farmer, *Infections and Inequalities: The Modern Plagues* (Berkeley: University of California Press, 1999); John Nott, "'No One May

Starve in the British Empire': Kwashiorkor, Protein and the Politics of Nutrition between Britain and Africa," *Social History of Medicine* 34, no. 2 (May 2021): 553–576; Marilyn Little, "Imperialism, Colonialism, and the New Science of Nutrition: The Tanganyika Experience," *Social Science and Medicine* 32, no. 1 (1991): 11–14.

24. See, for example, Ken DeBevoise, *Agents of Apocalypse: Epidemic Diseases in the Colonial Philippines* (Princeton, NJ: Princeton University Press, 1995); Randall Packard, *The Making of a Tropical Disease: A Short History of Malaria*, 2nd ed. (Baltimore, MD: Johns Hopkins University Press, 2021); Warwick Anderson, *Colonial Pathologies: American Tropical Medicine, Race, and Hygiene in the Philippines* (Durham, NC: Duke University Press, 2006).

25. Alexandre I. R. White, *Epidemic Orientalism: Race, Capital and the Governance of Infectious Disease* (Stanford, CA: Stanford University Press, 2023).

26. Christian McMillen, *Discovering Tuberculosis: A Global History, 1900 to the Present* (New Haven, CT: Yale University Press, 2015), 52–55.

27. John Nott, "'How Little Progress?' A Political Economy of Postcolonial Nutrition," *Population and Development Review* 44, no. 4 (December 2018): 777.

28. Robert A. Aronowitz, *Risky Medicine: Our Quest to Cure Fear and Uncertainty* (Chicago: The University of Chicago Press, 2015); Jeremy A. Greene, *Prescribing by Numbers: Drugs and the Definition of Disease* (Baltimore, MD: Johns Hopkins University Press, 2006).

29. Examples of studies of imperialism, food, and health in South Asia can be found in the "Suggested Further Reading" section at the back of this book.

30. Ruth Rogaski, *Hygienic Modernity: Meanings of Health and Disease in Treaty-Port China* (Berkeley: University of California Press, 2004).

31. On changes in East Asian cuisine over time and the development of ideas about national and regional Chinese cuisines, see "Suggested Further Reading."

32. Roel Sterckx, ed., *Of Tripod and Palate: Food, Politics, and Religion in Traditional China* (New York: Palgrave Macmillan, 2005); Roel Sterckx, *Food, Sacrifice, and Sagehood in Early China* (Cambridge: Cambridge University Press, 2011).

33. For a recent summary of the state of scholarship on this topic, see Hilary A. Smith, "Food, Health, and Nutrition in Chinese History," *History Compass* 20, no. 1 (January 2022): 1–12. Other studies of food and health in East Asia can be found in "Suggested Further Reading." Health is addressed in a few sections or chapters of the following: E. N. Anderson, *Food and Environment in Early and Medieval China* (Philadelphia: University of Pennsylvania Press, 2014); Geoffrey Barstow, *Food of Sinful Demons: Meat, Vegetarianism, and the Limits of Buddhism in Tibet* (New York: Columbia University Press, 2018); Jia-Chen Fu, *The Other Milk: Reinventing Soy in Republican China* (Seattle: University of Washington Press, 2018); Seung-joon Lee, *Gourmets in the Land of Famine: The Culture and Politics of Rice in Modern Canton* (Stanford, CA: Stanford University Press, 2011).

34. On the development of nutritional paradigms in Europe and the United States, see "Suggested Further Reading."

35. Gyorgy Scrinis, *Nutritionism: The Science and Politics of Dietary Advice* (New York: Columbia University Press, 2013).

36. See "Suggested Further Reading" for work on the health effects of particular types of food and on nutrition-deficiency disorders.

37. "Food as Medicine, Medicine as Food" issue of *Journal of the History of Medicine and Allied Sciences* 73, no. 2 (April 2018).

38. "Food Matters" issue of *Osiris* 35, no. 1 (2020).

39. Elizabeth Neswald, David F. Smith, and Ulrike Thomas, eds., *Setting Nutritional Standards: Theory, Policies, Practices* (Rochester, NY: University of Rochester Press, 2017).

40. Bridie Andrews, *The Making of Modern Chinese Medicine, 1850–1960* (Honolulu: University of Hawai'i Press, 2015); Sean Hsiang-lin Lei, *Neither Donkey nor Horse: Medicine in the Struggle over China's Modernity* (Chicago: University of Chicago Press, 2014); Volker Scheid, *Chinese Medicine in Contemporary China: Plurality and Synthesis* (Durham, NC: Duke University Press, 2002); Eric I. Karchmer, *Prescriptions for Virtuosity: The Postcolonial Struggle of Chinese Medicine* (New York: Fordham University Press, 2022).

41. Dipesh Chakrabarty, *Provincializing Europe: Postcolonial Thought and Historical Difference*, 2nd ed. (Princeton, NJ: Princeton University Press, 2009), 16.

42. Nick Cullather, "The Foreign Policy of the Calorie," *American Historical Review* 112, no. 2 (2007), 337–364.

43. Scrinis, *Nutritionism.*

44. Cullather, "The Foreign Policy of the Calorie."

45. Hannah LeBlanc, "The Protein Gap: The Rise and Fall of a Charismatic Nutrient in International Public Health," *Bulletin of the History of Medicine* 97, no. 4 (Winter 2023): 589.

Chapter 1. The Standard Man and the Chinese Normal

1. William H. Adolph, "A 4000-Year Food Experiment," *Scientific American* 143, no. 6 (December 1930): 428.

2. Francis G. Benedict, *Vital Energetics: A Study in Comparative Basal Metabolism* (Washington, DC: Carnegie Institution of Washington, 1938), 92.

3. I. N. Love, "Interruptions in Metabolism," *JAMA* 29, no. 9 (August 28, 1897): 423–424.

4. William Beaumont, *Experiments and Observations on the Gastric Juice, and the Physiology of Digestion* (Edinburgh: Maclachlan & Stewart, 1838).

5. Graham Lusk, "A Respiration Calorimeter for the Study of Disease," *Archives of Internal Medicine* 15 no. 5, part 2 (1915): 793–804.

6. This usage is technically incorrect, since a food calorie is actually a thousand times larger than what a physicist would call a calorie. But when most people talk about food, they use the term calorie rather than kCal, or food calorie. In this book "calorie" consistently refers to the larger unit.

7. E. W. H. Cruikshank, *Food and Nutrition: The Physiological Bases of Human Nutrition* (Baltimore, MD: The Williams & Wilkins Company, 1946), 49.

8. W. O. Atwater, C. D. Woods, and F. G. Benedict, *Report of Preliminary Investigations on the Metabolism of Nitrogen and Carbon in the Human Organism, with a Respiration Calorimeter of Special Construction* (Washington, DC: Government Printing Office, 1897); J. A. Riche and G. F. Soderstrom, "The Respiration Calorimeter of the Russell Sage Institute of Pathology in Bellevue Hospital," *Archives of Internal Medicine* 15, no. 5 part 2 (1915): 805–828.

9. "The Basal Metabolism of Man," *JAMA* 63, no. 16 (1914): 1395–1396.

10. Elizabeth Neswald, "Strategies of International Community-Building in Early Twentieth-Century Metabolism Research: The Foreign Laboratory Visits of Francis Gano Benedict," *Historical Studies in the Natural Sciences* 43, no. 1 (February 2013): 1–40.

11. Francis Gano Benedict Papers, Harvard Library. Internet Archive URL: https://web.archive.org/web/20240214122929/https://hollisarchives.lib.harvard.edu/repositories/14/resources/4576.

12. Eugene F. DuBois, *Basal Metabolism in Health and Disease*, 2nd ed. (Philadelphia: Lea & Febiger, 1927), 165.

13. Francis G. Benedict, "Basal Metabolism: The Modern Measure of Vital Activity," *Scientific Monthly* 27, no. 1 (July 1928): 5–27.

14. Kevin Donnelly, *Adolphe Quetelet, Social Physics and the Average Men of Science, 1796–1874* (New York: Routledge, 2016).

15. Lyon Playfair, *On the Food of Man in Relation to His Useful Work* (Edinburgh: Edmonston and Douglas, 1865), 48.

16. Playfair, *On the Food of Man*, 53.

17. For example, in A. L. Benedict, *Golden Rules of Dietetics* (St. Louis, MO: C. V. Mosby Medical Book and Publishing, 1908), 16; J. Alan Murray, *The Economy of Food: A Popular Treatise on Nutrition, Food and Diet* (New York: D. Appleton and Company, 1911), 207.

18. "No Standard Man," *California and Western Medicine* 22, no. 5 (May 1924): 234.

19. J. W. F., "Man as Found in the Drug Store," *Eclectic Review*12, no. 3 (March 15, 1909): 65–67.

20. "Inter-Allied Scientific Food Commission," *The Hospital: The Workers' Newspaper of Administrative Medicine and Institutional Life, Administration, National Insurance and Health* 54, no. 1673 (June 29, 1918): 273–274.

21. Food and Agriculture Organization of the United Nations, *Calorie Requirements: Report of the Committee on Calorie Requirements* (Washington, DC, June 1950).

22. J. Arthur Harris and Francis G. Benedict, "Biometric Standards for Energy Requirements in Human Nutrition," *Scientific Monthly* 8, no. 5 (May 1919), 385–402.

23. Itai Bendavid et al., "The Centenary of the Harris-Benedict Equations: How to Assess Energy Requirements Best? Recommendations from the ESPEN Expert Group," *Clinical Nutrition* 40, no. 3 (2021): 698.

24. See Harris and Benedict, "Biometric Standards for Energy Requirements" and J. Arthur Harris and Francis G. Benedict, *A Biometric Study of Basal Metabolism in Man* (Washington, DC: Carnegie Institution of Washington, 1919).

25. Harris and Benedict, "A Biometric Study of Human Basal Metabolism," *Proceedings of the National Academy of Sciences of the United States of America*, December 1, 1918, 37.

26. Harris and Benedict, "Biometric Standards for Energy Requirements," 399; Harris and Benedict, *A Biometric Study of Basal Metabolism in Man*, 32–33.

27. H. G. Earle and J. Strickland Goodall, "Basal Metabolism and Its Clinical Measurement," *Lancet* 197, no. 5095 (April 23, 1921): 853–854; Francis G. Benedict, Lan-Chen Kung, and Stanley D. Wilson, "The Basal Metabolism and Urinary Nitrogen Excretion of Chinese, Manchus and Others of the Mongolian Race," *Chinese Journal of Physiology* 12, no. 1 (1937): 67–100.

28. Grace MacLeod, Elizabeth E. Crofts, and Francis G. Benedict, "The Racial Factor in Metabolism," *Proceedings of the National Academy of Sciences of the United States of America*, June 1, 1925, 343.

29. Bendavid et al., "The Centenary of the Harris-Benedict Equations," 691.

30. D. F. Roberts, "Basal Metabolism, Race, and Climate," *Journal of the Royal Anthropological Institute of Great Britain and Ireland* 82, no. 2 (July–December 1952): 169.

31. MacLeod et al., "The Racial Factor in Metabolism"; Edward L. Turner, "The Racial Factor in Basal Metabolism," *JAMA* 87, no. 25 (December 1926): 2052–2053; "Racial and Dietary Factors in Basal Metabolism," *JAMA* 99, no. 20 (1932): 1692–1693; Francis G. Benedict, "The Racial Element in Human Metabolism," *American Journal of Physical Anthropology* 16, no. 4 (April–June 1932): 463–473; Francis G. Benedict, "Race: A Factor in Human Metabolism," *Proceedings of the American Philosophical Society* 78, no.1 (October 22, 1937): 101–110.

32. According to a summary of the September 7, 1926, Special Meeting of the Chinese Physiological Society published at the back of the *Chinese Journal of Physiology* 1, no. 4 (1927): 455.

33. John H. Foster and P. L. Hsieh, "The Vital Capacity of the Chinese: An Occupational Study," *Archives of Internal Medicine* 32, no. 3 (1923): 335–342; Grace MacLeod, Elizabeth E. Crofts, and Francis G. Benedict, "The Basal Metabolism of Some Orientals," *American Journal of Physiology* 73, no. 2 (July 1925): 449–462; Turner, "The Racial Factor in Basal Metabolism."

34. They used the Harris-Benedict, Aub-DuBois, and Dreyer formulas for their predictions.

35. MacLeod et al., "The Basal Metabolism of Some Orientals," 461.

36. A. R. Ruis, "'Children with Half-Starved Bodies' and the Assessment of Malnutrition in the United States, 1890–1950," *Bulletin of the History of Medicine* 87, no. 3 (Fall 2013): 378–406.

37. MacLeod et al., "The Basal Metabolism of Some Orientals," 459.

38. MacLeod et al., "The Basal Metabolism of Some Orientals," 462.

39. MacLeod et al., "The Basal Metabolism of Some Orientals," 450.

40. Roberts, "Basal Metabolism, Race, and Climate," 175.

41. Earle's article as described in MacLeod et al., "The Basal Metabolism of Some Orientals," 451.

42. MacLeod et al., "The Basal Metabolism of Some Orientals," 449.

43. H. Necheles, "Basal Metabolism in Orientals," *American Journal of Physiology* 91, no. 2 (January 1930): 661.

44. Nathan Hopson, "Nutrition as National Defense: Japan's Imperial Government Institute for Nutrition, 1920–1940," *Journal of Japanese Studies* 45, no. 1 (2019): 1–29.

45. Seizaburo Okada, Eiichi Sakurai, Toshio Kameda., "The Basal Metabolism of the Japanese," *Archives of Internal Medicine* 38, no. 5 (November 1926): 591.

46. Okada et al., "The Basal Metabolism of the Japanese," 601.

47. Hideo Takahira, "The Basal Metabolism of Normal Japanese Men and Women," in *Progress of the Science of Nutrition in Japan*, edited by Tadasu Saiki, 11–36 (Geneva: League of Nations Health Organization, 1926).

48. Okada et al., "The Basal Metabolism of the Japanese"; Yoichi Hatamoto, Kazuko Ishikawa-Takata, and Shigeho Tanaka, "History of Studies on Energy Requirements and Anthropometry in Japanese at the National Institute of Health and Nutrition," *Japanese Journal of Nutrition and Dietetics*, 78, Supplement (2000): S71-S79; Takahira, "The Basal Metabolism of Normal Japanese Men and Women."

49. DuBois, *Basal Metabolism in Health and Disease*, 150.

50. William H. Adolph, "Aspects of Nutrition and Metabolism in China," *Scientific Monthly* 29, no.1 (July 1929): 39–43.

51. M. Damodaran and K. G. Ananthanarayanan, "Is There a Racial Factor in Metabolism?" *Current Science* 4, no. 2 (August 1935), 96–97.

52. Chinese American Museum of Chicago, "1921: Birds Nests & the First Chinese Woman PhD in Chicago." Internet Archive URL: https://web.archive.org/web/20240703160032/https://ccamuseum.org/1921-birds-nests-the-first-chinese-woman-phd-in-chicago/.

53. Chi Che Wang and Jean E. Hawks, "Basal Metabolism of Twenty-One Chinese Children Reared or Born and Reared in the United States," *American Journal of Diseases of Children* 44, no.1 (1932): 79.

54. Chi Che Wang, "Basal Metabolism of American-Born Chinese Girls and of American Girls of the Same Age," *JAMA* 48, no.5 (1934): 1041–1049.

55. Wang, "Basal Metabolism of American-Born Chinese Girls," 1049.

56. Wang, "Basal Metabolism of American-Born Chinese Girls," 1049.

57. H. G. Earle, "Basal Metabolism of Chinese and Westerners," *Chinese Journal of Physiology, Report Series* 1 (1928): 59–92.

58. Earle, "Basal Metabolism of Chinese and Westerners."

59. Francis G. Benedict and H. S. D. Garven, "The Basal Metabolism of Male Chinese in Manchuria," *Chinese Journal of Physiology* 10, no. 1 (1936): 141–148; Benedict et al., "The Basal Metabolism and Urinary Nitrogen Excretion of Chinese, Manchus and Others of the Mongolian Race."

60. Leslie G. Kilborn and Francis G. Benedict, "The Basal Metabolism of Chinese in Szechwan," *Chinese Journal of Physiology* 11, no. 1 (137): 107–126.

61. At this time, the city we now know as Beijing was called Beiping.

62. Kilborn and Benedict, "The Basal Metabolism of Chinese in Szechwan."

63. Benedict, "Race: A Factor in Human Metabolism," 101.

64. Benedict, "Race: A Factor in Human Metabolism," 110.

65. Harris and Benedict, "Biometric Standards for Energy Requirements," 385.

66. H. Takahira et al., "Caloric Requirement of the Japanese People" (Tokyo: Imperial Governmental Institute for Nutrition, 1924), 16.

67. George Solt, *The Untold History of Ramen: How Political Crisis in Japan Spawned a Global Food Craze* (Berkeley: University of California Press, 2014), 49–55.

68. S. Suzuki, "Basal Metabolism in the Japanese Population," in *World Review of Nutrition and Dietetics*, edited by Geoffrey H. Bourne (New York: S. Karger, 1960), 122.

69. Food and Agriculture Organization, *Calorie Requirements*, 1.

70. Hsien Wu had by then become a permanent resident of the United States.

71. Food and Agriculture Organization, *Calorie Requirements*, 10–11.

72. Food and Agriculture Organization, *Calorie Requirements*, 8.

73. James Trefil and Margaret Hindle Hazen, *Good Seeing: A Century of Science at the Carnegie Institution of Washington, 1902–2002* (Washington, DC: Joseph Henry Press, 2002), 67.

74. In her article reviewing changes in the FAO's energy-requirement recommendations, Anna Ferro-Luzzi notes that even the 1996 committee encountered an "almost total lack of data [about basal metabolic rates] from developing countries" ("The Conceptual Framework for Estimating Food Energy Requirement," *Public Health Nutrition* 8, no. 7A [October 2005]: 945).

75. Ferro-Luzzi, "The Conceptual Framework"; Food and Agriculture Organization of the United Nations, World Health Organization & United Nations University, *Energy and Protein Requirements* (Geneva: World Health Organization, 1985).

76. Food and Agriculture Organization, *Energy and Protein Requirements*, 36–37.

77. C. J. K. Henry, "Basal Metabolic Rate Studies in Humans: Measurement and Development of New Equations," *Public Health Nutrition* 8, no. 7A (October 2005): 1135.

78. Camps et al., "Estimation of Basal Metabolic Rate in Chinese: Are the Current Prediction Equations Applicable?" *Nutrition Journal* 15, no. 79 (2016): 1–8; Keiko Ono Case, "Resting Energy Expenditures in Asian Women Measured by Indirect Calorimetry Are Lower than Expenditures Calculated from Prediction Equations," *Journal of the American Dietetic Association* 97, no. 11 (November 1997): 1288–1292; X. Yang et al., "Basal Energy Expenditure in Southern Chinese Healthy Adults: Measurement and Development of a New Equation," *British Journal of Nutrition* 104, no. 12 (2010): 1817–1823; H. Y. Liu, Y. F. Lu, and W. J. Chen, "Predictive Equations for Basal Metabolic Rate in Chinese Adults: A Cross-Validation Study," *Journal of the American Dietetic Association.* 95, no. 12 (1995): 1403–1408.

79. "Precision Nutrition: What Is It?" *Nutrition Source*, Harvard T. H. Chan School of Public Health, 2026. Internet Archive URL: https://web.archive.org/web/20201003200938/https://www.hsph.harvard.edu/nutritionsource/precision-nutrition/.

80. "Nutrition for Precision Health," Common Fund, National Institutes of Health, November 17, 2025. Internet Archive URL: https://web.archive.org/web/20250620143931/https://commonfund.nih.gov/nutritionforprecisionhealth.

Chapter 2. The Deficient Staple

1. M. E. Jaffa, *Nutrition Investigations among Fruitarians and Chinese at the California Agricultural Experiment Station, 1899–1901* (Washington, DC: Government Printing Office, 1901), 40. For more on Jaffa, see Patricia B. Swan and Kenneth J. Carpenter, "Myer E. Jaffa: Pioneering Chemist in the Food and Nutrition Sciences," *Bulletin for the History of Chemistry* 21 (1998): 51–57.

2. Jaffa, *Nutrition Investigations*, 25.

3. Jaffa, *Nutrition Investigations*, 36.

4. American Federation of Labor, "Some Reasons for Chinese Exclusion. Meat vs. Rice. American Manhood against Asiatic Coolieism. Which Shall Survive?" (Washington, DC: Government Printing Office, 1902): 24.

5. Francesca Bray, "Lord Millet in Alibaba's Cave: The Resurrection of an Iconic Food," in *Modern Chinese Foodways*, edited by Jia-Chen Fu, Michelle T. King, and Jakob A. Klein, 25–48 (Cambridge, MA: MIT Press, 2025); K. C. Chang, "Ancient China," in *Food in Chinese*

Culture: Anthropological and Historical Perspectives, edited by K. C. Chang, 26–27, 42–44 (New Haven, CT: Yale University Press, 1977); Ying-shih Yü, "Han," in *Food in Chinese Culture: Anthropological and Historical Perspectives*, edited by K. C. Chang, 70–73 (New Haven, CT: Yale University Press, 1977).

6. Zeng Xiongsheng 曾雄生, *Zhongguo daoshi yanjiu* 中国稻史研究 [Research on the history of rice in China] (Beijing: Zhongguo nongye chubanshe 中国农业出版社, 2018).

7. He Kejian 何克諫, *Zeng zhu bei zai shiwu bencao* 增註備載食物本草 [Expanded, annotated, and detailed *materia dietetica* (韓文堂, 1732), 9a–10a, as reprinted in Xie Weiqiang 謝偉強, ed., *Shiwu bencao dian ping* 食物本草點評 [Punctuated *materia dietetica* with commentary], with commentary by Jiang Runxiang 江潤祥 (Hong Kong: Cema wenchuang you xian gongsi 策馬文創有限公司, 2012).

8. He, *Zeng zhu bei zai shiwu bencao*, 9a–10a. The term that classical Chinese medical texts generally use to describe visceral systems of function—what we might call organs—is *zangfu* 臟腑. Within that set, generally consisting of eleven or twelve items, depending on the text, *zang* organs are distinguished from *fu* organs. Modern textbooks of Traditional Chinese Medicine (TCM) have attributed different kinds of functions to the two types, with *zang* organs being relatively *yin* on the yin-yang spectrum, while the *fu* organs were relatively *yang*. According to Nathan Sivin's analysis of modern TCM, the *zang* organs "control and regulate vital functions, while the yang [i.e., *fu*] systems receive, digest, store, transmit, and excrete vital substances and residues." See Nathan Sivin, *Traditional Medicine in Contemporary China* (Ann Arbor, MI: Center for Chinese Studies, 1987), 131.

9. Erik Millstone and Tim Lang, *The Atlas of Food: Who Eats What, Where and Why* (Berkeley: University of California Press, 2013).

10. Tom Jaine, ed., *The Oxford Companion to Food* (Oxford: Oxford University Press, 2014).

11. *Oxford English Dictionary*, 2nd ed. (Oxford: Oxford University Press,1989), under "staple."

12. "The Land of Potatoes," *The Western Herald* (Steubenville, OH) IX, no. 16 (April 26, 1816): 2, italics in original. Untitled note from *the Hampden Federalist and Public Journal* VXII, no. 22 (May 29, 1822): 87.

13. William Beckford, *The School for Good Living; Or, a Literary and Historical Essay on the European Kitchen: Beginning with Cadmus the Cook and King, and Concluding with the Union of Cookery and Chymistry* (London: J. Gillet, 1814), 9.

14. Douglass H. Thomson, Jack G. Voller, and Frederick S. Frank, eds., *Gothic Writers: A Critical and Bibliographical Guide* (Westport, CT: Greenwood Press, 2002): 54–55.

15. Beckford, *The School for Good Living*, 228–229.

16. For an example of American concerns about "rice-eating" Chinese immigrants as competitors in the labor market, see M. C. Briggs, "The Chinese—Must They Go? FACTS AND FABLES," *Christian Advocate* 54, no. 13 (March 27, 1879): 193.

17. "Wheat as a Civilizer," *American Farmer, a Monthly Magazine of Agriculture and Horticulture* 1, no. 3 (March 1870): 39.

18. G. M. Beard, "The Diet of Brain-Workers," *Hours at Home: A Popular Monthly of Instruction and Recreation* 9, no. 5 (September 1869): 423.

19. Philip A. Kuhn, *Chinese among Others: Emigration in Modern Times* (New York: Rowman & Littlefield, 2008).

20. Although it may seem surprising to coin a term for the disparaged Other before coining one for the valued Self, this is in fact the usual sequence. The Self is taken as natural and obvious, not requiring a special label, so it is only after the naming of the quality or behavior perceived as aberrant that the quality or behavior perceived as normal acquires a name. Thus, for example, "homosexual" was coined before "heterosexual."

21. "Wheat as a Civilizer," 39.

22. William Crookes, "Address of the President before the British Association for the Advancement of Science," *Science* New Series 8, no. 200 (October 28, 1898): 564.

23. Crookes, "Address of the President," 569.

24. To arrive at these numbers, I added together the reported number of studies for each category that was either clearly marked as consisting of northern Europeans or Americans of northern European descent (such as "Swedish mechanics") or that we can assume, based on the typical ethnic makeup at this time of the kind of group specified, were mostly northern Europeans or Americans of northern European descent (such as "College clubs in the United States" or "Rowing clubs in New England"). I have not included in the calculations categories that were more ethnically ambiguous, such as "Laborers' families in large cities of United States." The 72 different studies of northern Europeans and Americans of northern European descent represent just the categories that are clearly marked, so that is why I describe 72 as the minimum number of studies with such a focus.

25. Hermann T. Vulté and Sadie B. Vanderbilt, *Food Industries: An Elementary Text Book on the Production and Manufacture of Staple Foods*, 2nd ed. (Easton, PA: The Chemical Publishing Company, 1916), 45.

26. G. S. Fraps, *The Composition of Rice and Its By-Products* (College Station, TX: Texas Agricultural Experiment Station, 1916), 17.

27. Elmer Verner McCollum, *The Newer Knowledge of Nutrition: The Use of Food for the Preservation of Vitality and Health* (New York: Macmillan, 1919).

28. The etymology of the word "beriberi" is uncertain, despite confident-sounding pronouncements in various dictionaries. For a discussion of its etymology, see Hilary A. Smith, *Forgotten Disease: Illnesses Transformed in Chinese Medicine* (Stanford, CA: Stanford University Press, 2017), 118–120.

29. The original term combined "vital" and "amines"—"vital" because these elements were necessary to sustain life, and "amines" because Casimir Funk, the scientist who coined the term, assumed that the elements were amines in their chemical structure. When scientists began to elucidate the chemical structure of the accessory factors, however, it turned out that they were not all amines, so the word was shortened to "vitamins." Chemists often used the "-in" ending to indicate "a neutral substance of undefined composition," so it fit perfectly the loosely defined category that vitamins represented. See Jack Cecil Drummond, "The Nomenclature of the So-Called Accessory Food Factors," *Biochemical Journal* 14, no. 5 (1920): 660.

30. "Final Report of the Mixed Committee of the League of Nations on the Relation of Nutrition to Health, Agriculture and Economic Policy" (Geneva: 1937), 318. See also Hilary A. Smith, "Absorbing Vitamins: How a Nutritional Paradigm Was Reinvented in Republican China," in *Modern Chinese Foodways*, edited by Jia-Chen Fu, Michelle T. King, and Jakob A. Klein, 89–108 (Cambridge, MA: MIT Press, 2025).

31. According to Seung-joon Lee, the first steam-powered rice mill in southeast Asia opened in Bangkok in 1858, and by the late 1800s ethnic Chinese businesspeople dominated the rice trade in the region at all stages: buying it from producers, milling it, selling it, and distributing it. Lee, "Rice and Maritime Modernity: The Modern Chinese State and the South China Sea Rice Trade," in *Rice: Global Networks and New Histories*, edited by Francesca Bray, Peter A. Coclanis, and Dagmar Schäfer, 99–117 (Cambridge: Cambridge University Press, 2015).

32. J. H. DeForest, "The Difference between Japan and the United States," *The Independent* 60, no. 2999 (May 24, 1906): 1199.

33. For "cradle of disease," see Ari L. N. Heinrich, *The Afterlife of Images: Translating the Pathological Body between China and the West* (Durham, NC: Duke University Press, 2008). For "land of famine," see Walter H. Mallory, *China: Land of Famine* (New York: American Geographical Society, 1928). For "Sick man of Asia," see Marta Hanson, "From Sick Man of Asia to Sick Uncle Sam," *Current History* 119, no. 818 (2020): 241–244.

34. Robert McCarrison, "Nutrition and National Health, Lecture I: Food, Nutrition, and Health," *Journal of the Royal Society of Arts* 84, no. 4371 (August 28, 1936): 1056, 1058.

35. Elmer Verner McCollum, *Lectures on Nutrition: A Series of Lectures Given at the Mayo Foundation and the Universities of Wisconsin, Nebraska, Iowa, and Washington (St. Louis), 1924–1925* (Philadelphia: W. B. Saunders Company, 1925), 165.

36. Zheng Zhenwen 鄭貞文, *Yingyang huaxue* 營養化學 [Nutritional chemistry] (Shanghai: Commercial Press, 1929).

37. Wu Xian 吳憲 (Hsien Wu), "Wo guoren zhi chifan wenti" 我國人之吃飯問題" [My countrymen's eating problem], *Duli pinglun* 獨立評論 [Independent commentary] no. 2 (1932): 14–18; "Zhongguoren de yingyang wenti 中國人的營養問題" [Chinese people's nutrition problem], *Ziran jie* 自然界 [Natural world] 5, no. 1 (1930): 28–36; "Zhongguo ren de yingyang wenti 中國人的營養問題" [Chinese people's nutrition problem], *Tu wei* 突圍 [Breaking the siege] 1, no. 7 (1934): 7–8; Chen Yuxin 陳育信, "Guomin yingyang wenti 國民營養問題" [Chinese citizens' nutrition problem], *Qing nian Zhongguo* 青年中國 [Young China] 1, no. 12 (1937): 9–11.

38. Liu Ruiheng 劉瑞恆, "Yingyang wenti 營養問題" [Nutrition problems], *Gonggong weisheng yuekan* 公共衛生月刊 [Public health monthly] 2, no. 5 (1936): 346.

39. Liu, "Yingyang wenti," 346.

40. Hsien Wu (Wu Xian 吳憲), "Nutritional Deficiencies in China and Southeast Asia," in *Proceedings of the Fourth International Congress on Tropical Medicine and Malaria*, vol. 2 (Washington, DC: May 10–18, 1948), 1218, 1219, 1222.

41. Wu Xian 吳憲 (Hsien Wu), *Yingyang gailun* 營養概論 [Introduction to nutrition], 6th ed. (Shanghai: Commercial Press, 1947), 12. See also Zheng Ji 鄭集 (Libin T. Cheng), *Shiyong yingyangxue* 使用營養學 [Applied nutrition] (Nanjing: Zheng Zhong shu ju, 1947).

42. H. C. Hou et al., *Nutritional Studies in Shanghai* (Shanghai: Henry Lester Institute of Medical Research, 1940), 83.

43. Zheng, *Yingyang huaxue*, 32.

44. Xu Wancheng 許晚成, "Maifan yu mifan zhi bijiao 麥飯與米飯之比較" [A comparison of wheat and rice], *Shun pao* 申報 (English title *Shanghai News*), October 27, 1925), 5.

45. Zheng, *Yingyang huaxue*, 12.

46. Chi-Shih Yang and K.K. Huang, "Beri-beri in Nanking," *Chinese Medical Journal* 48 (1934): 20.

47. Qin Daojian 秦道堅, "Gailiang cao mi 改良糙米" [Improving coarse rice], *Jiating yiyao* 家庭醫藥 [Household medicine] no. 2 (October 1, 1946): 42–43.

48. Food and Agriculture Organization, "Report of the Rice Study Group" (Washington, DC: July 1947).

49. Literature on these famines can be found in "Suggested Further Reading."

50. Food and Agriculture Organization, "Report of the Rice Study Group," 3.

51. Food and Agriculture Organization, "Report of the Rice Study Group," 5.

52. Food and Agriculture Organization, "Report of the Rice Study Group," 57, 56.

53. R. Passmore, "Obituary Notice: Wallace Ruddell Aykroyd, CBE, MD, ScD," *British Journal of Nutrition* 43, no. 2 (1980): 245–250.

54. W. R. Aykroyd, "Malnutrition and the Rice Problem," *Proceedings of the Fourth International Congress on Tropical Medicine and Malaria*, vol. 2 (Washington, DC: May 10–18, 1948), 1172.

55. Aykroyd, "Malnutrition and the Rice Problem," 1173.

56. George Solt, *The Untold History of Ramen: How Political Crisis in Japan Spawned a Global Food Craze* (Berkeley: University of California Press, 2014).

57. Vaclav Smil and Kazuhiko Kobayashi, *Japan's Dietary Transition and Its Impacts* (Cambridge, MA: MIT Press, 2012).

58. Both quotations found in Solt, *The Untold History of Ramen*, 78–79.

59. Barry Riley, *The Political History of American Food Aid: An Uneasy Benevolence* (Oxford: Oxford University Press, 2017), 168–192.

60. "Nongshi xiaozu zhang huiyi jilu 農事小組長會議記錄" [Meeting minutes of the head of the Agricultural Affairs Small Group], in *Shi guang, shi guang—Taiwan yinshi dang'an gushi* 食光· 時光— 台灣飲食檔案故事 [Eras and the glory of food—Stories from Taiwan's food archives], edited by Chen Yixing 陳怡行 (Hsin bei: National Development Committee Archive Management Bureau/Ministry of Culture/Hakka Committee Hakka Cultural Development Center, 2021), 58.

61. "Yuangong buxi jiangxi 員工補習講習" [Employees' supplementary study lectures], in *Shi guang, shi guang—Taiwan yinshi dang'an gushi* 食光· 時光— 台灣飲食檔案故事 [Eras and the glory of food—Stories from Taiwan's food archives], edited by Chen Yixing (Hsin bei: National Development Committee Archive Management Bureau/Ministry of Culture/Hakka Committee Hakka Cultural Development Center, 2021), 56.

62. Emphasis in original. "Nongshi xiaozu zhang huiyi jilu," 58.

63. Chen Yixing, ed., *Shi guang, shi guang—Taiwan yinshi dang'an gushi* 食光· 時光— 台灣飲食檔案故事 [Eras and the glory of food—Stories from Taiwan's food archives] (Hsin bei: National Development Committee Archive Management Bureau/Ministry of Culture/Hakka Committee Hakka Cultural Development Center, 2021), 53.

64. Michelle T. King, *Chop Fry Watch Learn: Fu Pei-mei and the Making of Modern Chinese Food* (New York: Norton, 2024): 71.

65. Chen Xinqian 陳新謙, "Yinggai gai chi biaozhun mimian 應該改吃標準米面" [We should change to eating standard rice and wheat], *Kexue dazhong* 科學大眾 [Popular science] no. 4 (1950): 121–123.

66. Chen, "Yinggai gai chi biaozhun mimian," 123.

67. Chen, "Yinggai gai chi biaozhun mimian."

68. Jin Daxun 金大勳 et al., "Zaliang, caoliang yu xiliang shi de biaoguan xiaohualü—I. cu xianwei, zhifang yu danbaizhi de Xiaohua 雜糧, 糙糧與細糧食的表觀消化率" [The apparent digestibility of mixed grains, unrefined grains and refined grains—I. Rough fiber, fat and protein digestion] *Yingyang xuebao* 营养学报 [Acta nutrimenta sinica] 1, no. 3 (September 1956): 207–218; Yang Tongshu 楊同書 and Hou Xiangchuan 侯祥川, "Shanghai shi 20 zhong mi zhong weishengsu B1 de hanliang 上海市20種米中維生素B_1的含量" [The vitamin B1 content of 20 kinds of rice in Shanghai], *Yingyang xuebao* 營養學報 [Acta nutrimenta sinica] 2, no. 3 (September 1957): 217–218; Shen Zhiping 沈治平 et al., "Daomi yingyang chengfen de yanjiu 稻米營養成分的研究" [Research on the nutritional constituents of paddy rice], *Yingyang xuebao* 营养学报 [Acta nutrimenta sinica] 3, no. 2 (June 1958): 70–75.

69. Yang Enfu 楊恩孚, "Wo guo youguan yingyang wenti yanjiu gongzuo de xianzhuang yu jianglai 我國有關營養問題研究工作的現狀與將來" [The status and future of China's research work concerning nutritional problems], *Shengli kexue jinzhan* 生理科學進展 [Progress in physiological sciences] 1, no. 1 (March 1957): 31–37; Li Zhongying 李鍾英, "'Jiu er' mi de yingyang chengfen「九二」米的營養成分" [The nutritional constituents of 'nine-two' rice], *Fujian zheng bao* 福建政報 [Fujian political report] no. 5 (1950): 65.

70. On the politics of Maoist science, see Sigrid Schmalzer, *Red Revolution, Green Revolution: Scientific Farming in Socialist China* (Chicago: University of Chicago Press, 2016); Darryl E. Brock and Chunjuan Nancy Wei, eds. *Mr. Science and Chairman Mao's Cultural Revolution: Science and Technology in Modern China* (Lanham, MD: Lexington Books, 2013).

71. Shellen X. Wu, "Yuan Longping (1930–2021)," *Nature* 595, no. 7865 (July 1, 2021): 26.

72. Schmalzer, *Red Revolution, Green Revolution*, chap. 3.

73. Schmalzer, *Red Revolution, Green Revolution*, 96–98.

74. "Bai mi fan—la ji shi pin zhi wang"「白米饭— 垃圾食品之王」[White rice—the king of junk food] *Ying kang she* 盈康社 [Yield Healthy Life (nutritional products company)] (January 11, 2022). Internet Archive URL: https://web.archive.org/web/20251018171249/https://www.yieldhealthylife.com/post/%E7%99%BD%E7%B1%B3%E9%A3%AF-%E5%9E%83%E5%9C%BE%E9%A3%9F%E5%93%81%E4%B9%8B%E7%8E%8B.

75. "The Global Rice Crisis," *The Economist*, April 1, 2023, 27–28.

Chapter 3. The Wrong Kind of Protein

1. Hsien Wu (Wu Xian 吳憲), "The Chinese Diet in the Light of Modern Knowledge of Nutrition," *Chinese Social and Political Science Review* 11, no. 1 (January 1927): 74.

2. The Boxer indemnity scholarships sent a total of about 1,300 Chinese students to the United States between 1909 and 1937. After eight foreign nations sent troops to China to put down the Boxer uprising in 1900, the Qing government was forced to pay a large indemnity to all of them. The Americans committed part of their indemnity to building US-style educational institutions in China and to funding Chinese students to study in the United States. Hsien Wu was among the promising young scholars sent abroad in 1911. See "The Boxer Indemnity Scholarship Program," earlychinesemit. Internet Archive URL: https://web.archive.org/web/20250826124337/https://earlychinesemit.mit.edu/three-waves/boxer-indemnity-scholarship-program/#_ftn2.

3. Wu Xian 吳憲 (Hsien Wu), "Zhongguo shiwu zhi xiandai yingyangxue shiguan 中國食物之現代營養學識觀" [Modern nutrition-science perspectives on Chinese foodways], *Yixue zhoukan ji* 醫學週刊集 [Medicine weekly] 1 (December 1927): 72–84. See Angela Ki Che Leung, "To Build or to Transform Vegetarian China: Two Republican Projects," in *Moral Foods: The Construction of Nutrition and Health in Modern Asia*, edited by Angela Ki Che Leung and Melissa L. Caldwell, 221–240 (Honolulu: University of Hawai'i Press, 2019).

4. William H. Adolph, "Diet Studies in Shantung," *China Medical Journal* 37 (1 December 1923): 1013–1019.

5. Zheng Ji 郑集 (Libin T. Cheng), *Zheng Ji kexue wen xuan* 郑集科学文选 [Selected scientific works of Zheng Ji] (Nanjing: Nanjing daxue chubanshe, 1993), 20.

6. Hsien Wu (Wu Xian 吳憲), "The Chinese Diet," 77.

7. Rachel Laudan, *Cuisine and Empire: Cooking in World History* (Berkeley: University of California Press, 2013), 256.

8. James Crichton-Browne, *Parcimony in Nutrition* (London: Funk and Wagnalls, 1909), 35.

9. Van Buren Denslow, *Principles of the Economic Philosophy of Society, Government and Industry* (New York: Cassell & Company, 1888), 402.

10. "RULERS OF THE WORLD," *Detroit Free Press*, February 4, 1903, 2.

11. Roel Sterckx, "Food and Philosophy in Early China," in *Of Tripod and Palate: Food, Politics, and Religion in Traditional China*, edited by Roel Sterckx, 34–61 (New York: Palgrave MacMillan, 2005).

12. Françoise Sabban, "La viande en Chine: Imaginaire et usages culinaires," *Anthropozoologica* no. 18 (1993): 82.

13. On hunting statutes, see E. N. Anderson, *Food and Environment in Early and Medieval China* (Philadelphia: University of Pennsylvania Press, 2014); on archaeological evidence, see K. C. Chang, "Ancient China," and Ying-shih Yü, "Han China," in *Food in Chinese Culture: Anthropological and Historical Perspectives*, edited by K. C. Chang (New Haven, CT: Yale University Press, 1977), 29 and 58.

14. Sabban, "La viande en Chine," 82.

15. Thomas David DuBois, "Counting the Carnivores: Who Ate Meat in Republican China?" *Social Science History* 46, no. 4 (Winter 2022): 769.

16. Zheng, *Zheng Ji kexue wen xuan*, 18.

17. John Kieschnick, "Buddhist Vegetarianism in China," in *Of Tripod and Palate: Food, Politics and Religion in Traditional China*, edited by Roel Sterckx, 186–212 (New York: Palgrave Macmillan, 2005).

18. Wang Xuetai 王学泰, *Zhongguo yinshi wenhua shi* 中国饮食文化史 [A history of Chinese food culture] (Beijing: Zhongguo qing nian chubanshe, 2012), 302–305; Kieschnick, "Buddhist Vegetarianism in China."

19. Eric Reinders, "Blessed Are the Meat Eaters: Christian Antivegetarianism and the Missionary Encounter with Chinese Buddhism," *positions: east asia cultures critique* 12, no. 2 (2004): 514, 521–523.

20. Wu Xian 吳憲 (Hsien Wu), *Yingyang gailun* 營養概論 [Introduction to nutrition], 6th ed. (Shanghai: Commercial Press, 1947), 86–87.

21. Laudan, *Cuisine and Empire*, Chapter 3.

22. Eric M. Greene, "A Reassessment of the Early History of Chinese Buddhist Vegetarianism," *Asia Major* Third Series 29, no. 1 (2016): 37.

23. Chen Zhiyuan 陳志遠, "Liang Wudi yu seng tuan sushi gaige—jiedu 'Duan jiu rou wen' 梁武帝與僧團素食改革— 解讀《斷酒肉文》" [Emperor Wu of Liang and the vegetarian reform of the clergy—Interpreting the "Discourse on Giving Up Meat and Alcohol"], *Zhonghua wen shi lun cong* 中華文史論叢 [Essays on Chinese literature and history] no. 3 (2013): 93–121.

24. Vincent Goossaert, "The Beef Taboo and the Sacrificial Structure of Late Imperial Chinese Society," in *Of Tripod and Palate: Food, Politics, and Religion in Traditional China*, edited by Roel Sterckx, 237–248 (New York: Palgrave Macmillan, 2005).

25. On fake-meat gastronomy, see Sabban, "La viande en Chine," 85–86.

26. Katarzyna Cwiertka, *Modern Japanese Cuisine: Food, Power and National Identity* (London: Reaktion Books, 2006), 33; Chelsey Wong, "Eating to Survive: The Overlooked Reason for Beef Consumption in Japan, 1974–1912," *Global Food History* 11, no. 3 (2024): 1–28.

27. Aya Majima, "Eating Meat, Seeking Modernity: Food and Imperialism in Late Nineteenth and Early Twentieth Century Japan," in *Critical Readings on Food in East Asia*, vol. 1, edited by Katarzyna J. Cwiertka, 111–134 (Leiden: Brill, 2013).

28. See the comments by W. D. Halliburton in "Over-Nutrition and Under-Nutrition, with Special Reference to Proteid Metabolism," *The Lancet*, September 15, 1906, 730. Alan Rosen, "Lafcadio Hearn and the Problem of Japanese Food," *Kumamoto University English Department Bulletin* 41, March 15, 1998, 26.

29. Hsien Wu, "The Chinese Diet," 78–79.

30. Warren J. Wilson, *Meat Makes People Powerful: A Global History of the Modern Era* (Iowa City: University of Iowa Press, 2018), chap. 1.

31. Wilson, *Meat Makes People Powerful*.

32. For more on the history of vegetarianism in Europe and America, see Tristram Stuart, *The Bloodless Revolution: A Cultural History of Vegetarianism from 1600 to Modern Times* (New York: Norton, 2007); Colin Spencer, *Vegetarianism: A History* (New York: Four Walls Eight Windows, 2002); Adam Shprintzen, *The Vegetarian Crusade: The Rise of an American Reform Movement, 1817–1921* (Chapel Hill: University of North Carolina Press, 2013); Karen Iacobbo, *Vegetarian America: A History* (Westport, CN: Praeger, 2004).

33. William E. A. Axon, "Some Chinese Arguments for Vegetarianism," *Good Health* 38, no. 4 (April 1903): 11–15.

34. John Smith, *Fruits and Farinacea, the Proper Food of Man; Being an Attempt to Prove, from History, Anatomy, Physiology and Chemistry that the Original, Natural, and Best Diet of Man Is Derived from the Vegetable Kingdom* (New York: Fowler and Wells, Philadelphia, 1856), 180.

35. Howard Williams, *The Ethics of Diet: A Catena of Authorities Deprecatory of the Practice of Flesh-Eating* (London: F. Pitman, 1883), 10.

36. R. Russell, *Strength and Diet: A Practical Treatise with Special Regard to the Life of Nations* (New York: Longmans, Green, & Co., 1905), 189–190.

37. Florence Peltier Perry, "The Vegetable Eating Nations," *Vegetarian Magazine* 4, no. 11 (August 1900): 3–4.

38. Mary Foote Henderson, *The Aristocracy of Health: A Study of Physical Culture, Our Favorite Poisons, and a National and International League for the Advancement of Physical Culture* (Washington, DC: Colton Publishing Company, 1904).

39. "Wu Ting Fang Returns to the United States," *Battle Creek Idea*, April 19, 1912.

40. Wu Tingfang, *America through the Spectacles of an Oriental Diplomat* (New York: Frederick A. Stokes Company, 1914), 196. See also Linda Pomerantz-Zhang, *Wu Tingfang (1842–1922): Reform and Modernization in Modern Chinese History* (Hong Kong: Hong Kong University Press, 1992).

41. Wu Tingfang, *America through the Spectacles of an Oriental Diplomat*, 210.

42. Jia-Chen Fu, *The Other Milk: Reinventing Soy in Republican China* (Seattle: University of Washington Press, 2018), Chapter 1.

43. "Shen shi weisheng hui 慎食衛生會" [Cautious Diet and Health Society], *Zhong Xi yixue bao* 中西醫學報 [Journal of Chinese and Western medicine] no. 6 (1910): 1.

44. Angela Ki Che Leung, "To Build or to Transform Vegetarian China." See also Lin Xi 林曦 and Li Yongchen 李永宸, "Shen shi weisheng hui yanjiu 慎食卫生会研究" [Study on the Cautious Diet and Health Society], *Zhonghua yishi zazhi* 中华医史杂志 [Chinese journal of the history of medicine] 50, no. 4 (July 2020): 200–206.

45. The Dutch chemist Gerardus Johannes Mulder is generally credited with coining the term, but archival research demonstrated that Berzelius actually came up with it. Hubert Bradford Vickery, "The Origin of the Word Protein," *Yale Journal of Biology and Medicine* 22, no. 5 (May 1950): 387–393. On the protein radical, see Kenneth J. Carpenter, *Protein and Energy: A Study of Changing Ideas in Nutrition* (Cambridge: Cambridge University Press, 1994), 43.

46. Hanne K. Maehre et al., "Protein Determination—Method Matters," *Foods* 7, no. 5 (2018): 1–11.

47. Kenneth J. Carpenter, "A Short History of Nutritional Science: Part I (1785–1885)," *Journal of Nutrition* 133, no. 3 (March 2003): 641.

48. Bee Wilson, "Protein Mania: The Rich World's New Diet Obsession," *The Guardian*, January 4, 2019. Internet Archive URL: https://web.archive.org/web/20260210143601/https://www.theguardian.com/news/2019/jan/04/protein-mania-the-rich-worlds-new-diet-obsession.

49. Thanks to Nathan Hopson for pointing me to the earliest known uses of 蛋白質 (in Japanese, *tanpaku*) in Japanese sources: Kawakami Kōmin's (川本幸民)'s 1861 *Kagaku shinsho* 化学新書 [New book of chemistry] and Shiba Ryōkai's (司馬凌海) 1862 *Shichi shin'yaku* 七新藥 [Seven new medicines]. Thanks, also, to J. J. Strange and Victor Mair for their sleuthing. Ge Keyou 葛可佑, chief ed., *Zhongguo yingyang kexue quan shu* 中国营养科学全书 [An overview of nutrition sciences in China], vol. 2 (Beijing: Renmin weisheng chubanshe, 2004), 30, confirms the German origin but omits the Japanese intermediary.

50. Lin and Li, "Shen shi weisheng hui yanjiu" 202.

51. Huiming Society, "Sushi zhuyi jue yi" 素食主意決疑 [Resolving some doubts about vegetarianism], *Zhong Xi yixue bao* 中西醫學報 [Journal of Chinese and Western medicine] 4, no. 3 (1913): 4–6.

52. Shou Yu, "Sushi zhuyi" 素食主意 [Vegetarianism], *Yu yu xin wen* 禹域新聞 [Land of Yu news], June 25, 1914, 2.

53. Carpenter, *Protein and Energy*, 107.

54. "The British Medical Association Meeting at Toronto," *The Lancet*, September 15, 1906, 730.

55. D. McCay, *The Protein Element in Nutrition* (London: Edward Arnold, 1912): 190.

56. McCay, *The Protein Element in Nutrition*, 206.

57. McCay, *The Protein Element in Nutrition*, 140.

58. John T. Edsall, "Hsien Wu and the First Theory of Protein Denaturation (1931)," *Advances in Protein Chemistry* 46 (1995): 1–5.

59. Wu Xian 吳憲 (Hsien Wu), "Danbaizhi lei zhi shengli de jiazhi 蛋白質類之生理的價值" [The physiological value of types of proteins], *Kexue* 科學 [Science] 11, no. 8 (1926): 1049–1050.

60. See Livia Kohn, *Daoist Dietetics: Food for Immortality* (St. Petersburg, FL: Three Pines Press, 2010).

61. Wu Xian, "Danbaizhi lei zhi shengli de jiazhi," 1049–1050. For comparison, the current daily recommended intake levels (DRIs) that Chinese nutrition scientists officially recommend are 75 to 90 g per day for men and 65 to 80 g per day for women, with light and heavy physical activity representing the two ends of the range. Ge Keyou, *Zhongguo yingyang kexue quan shu*, 2: 1778.

62. Fu, *The Other Milk*: 75–76 (comparing boys' heights), 85–86 (vegetarian rats), 85 (measured protein levels).

63. Wu Xian, "Danbaizhi lei zhi shengli de jiazhi," 1053.

64. Wu Xian, "Danbaizhi lei zhi shengli de jiazhi," 1052.

65. Wu Xian, *Yingyang gailun*, 56.

66. Hsien Wu, "The Chinese Diet," 68.

67. Leung, "To Build or to Transform Vegetarian China," 232.

68. Selected examples of this research include Lan-Chen Kung, H. L. Yeh, and W. H. Adolph, "The Availability of Calcium in Vegetable Food Materials," *Chinese Journal of Physiology* 13, no. 3 (1938): 307–316; Tung-Tou Chen, Hsien Wu, and Chang-Ying Chang, "Further Experiments with Soybean Diets," *Chinese Journal of Physiology* 16, no. 1 (1941): 91–96; Tung-Tou Chen, Chang-Ying Chang, and Tsung-Hsien Luo, "Cataract in Vegetarian Rats," *Chinese Journal of Physiology* 16, no. 2 (1941): 241–250; Chang-Ying Chang, Hsien Wu, and Tung-Tou Chen, "Life Span of Rats on Vegetarian and Omnivorous Diets," *Chinese Journal of Physiology* 16, no. 2 (1941): 229–240.

69. Wu Xian, *Yingyang gailun*, 138–139.

70. See Kathryn Jean Edgerton-Tarpley, "From 'Nourish the People' to 'Sacrifice the Nation': Changing Responses to Disaster in Late Imperial and Modern China," *Journal of Asian Studies* 73, no. 2 (2014): 447–469.

71. Carpenter, *Protein and Energy*, 157–160.

72. Carpenter, *Protein and Energy*, chap. 9; Warren Belasco, "Algae Burgers for a Hungry World? The Rise and Fall of Chlorella Cuisine," *Technology and Culture* 30, no. 3 (1997): 608–634; John Nott, "'How Little Progress?' A Political Economy of Postcolonial Nutrition," *Population and Development Review* 44, no. 4 (December 2018): 771–791; LeBlanc, "The Protein Gap."

73. Donald S. McLaren, "The Great Protein Fiasco," *The Lancet* 304, no. 7872 (July 13, 1974): 96.

74. McLaren, "The Great Protein Fiasco."

75. Carpenter, *Protein and Energy*, 160.

76. James E. Dalen et al., "The Epidemic of the 20th Century: Coronary Heart Disease," *American Journal of Medicine* 127, no. 9 (September 2014): 807–812.

Chapter 4. The White Milk Burden

1. Berthold Laufer, "Some Fundamental Ideas of Chinese Culture," *Journal of Race Development* 5, no. 2 (October 1914): 167.

2. Laufer, "Some Fundamental Ideas," 168.

3. Laufer, "Some Fundamental Ideas," 170.

4. See, for example, Anne Mendelson, *Spoiled: The Myth of Milk as Superfood* (New York: Columbia University Press, 2023); Anne Mendelson, *Milk: The Surprising Story of Milk through the Ages* (New York: Knopf, 2008).

5. Deborah Valenze, *Milk: A Global and Local History* (New Haven, CT: Yale University Press, 2011), 3; Peter Atkins, *Liquid Materialities: A History of Milk, Science and the Law* (Farnham: Ashgate, 2010), xv; E. Melanie DuPuis, *Nature's Perfect Food: How Milk Became America's Drink* (New York: New York University Press, 2002), 28.

6. Y. Itan et al., "The Origin of Lactase Persistence in Europe," *PLoS Computational Biology* 5, no. 8 (2009): e1000491; Andrew Curry, "Archaeology: The Milk Revolution," *Nature* 500, no.7460 (July 31, 2013): 20–22.

7. For an older version of my take on this history, see Hilary A. Smith, "Good Food, Bad Bodies: Lactose Intolerance and the Rise of Milk Culture in China," in *Moral Foods: The Construction of Nutrition and Health in Modern Asia*, edited by Angela Ki Che Leung and Melissa L. Caldwell, 262–284 (Honolulu: University of Hawai'i Press, 2019). Parts of that essay reappear in this chapter.

8. Wu Xian 吳憲 (Hsien Wu), *Yingyang gailun* 營養概論 [Introduction to nutrition], 6th ed. (Shanghai: Commercial Press, 1947), 99.

9. Frederick J. Simoons, "The Traditional Limits of Milking and Milk Use in Southern Asia," *Anthropos* 65, no. 3/4 (1970): 547–593, map 3; Andrea Wiley, "Milk for 'Growth': Global and Local Meanings of Milk Consumption in China, India, and the United States," *Food and Foodways* 19 (2011): 11–33.

10. H. T. Huang, "Hypolactasia and the Chinese Diet," *Current Anthropology* 43, no. 5 (December 2002): 816.

11. E. N. Anderson Jr. and Marja L. Anderson, "Modern China: South," in *Food in Chinese Culture: Anthropological and Historical Perspectives*, edited by K. C. Chang (New Haven, CT: Yale University Press, 1977), 341.

12. See also Huang, "Hypolactasia and the Chinese Diet."

13. *Huang Di nei jing su wen: An Annotated Translation of Huang Di's Inner Classic—Basic Questions*, vol. 1, translated by Paul U. Unschuld and Hermann Tessenow (Berkeley: University of California Press, 2011), 214–215.

14. H. T. Huang, "Fermentations and Food Science," in *Science and Civilisation in China*, vol. 6 *Biology and Biological Technology* (Cambridge: Cambridge University Press, 2000), 248.

15. Jia Sixie 賈思勰, *Qimin yaoshu* 齊民要術 [Essential techniques for the welfare of the people] (Taipei: Taiwan shangwu yinshuguan, 1968), 1:89.

16. Françoise Sabban, "Un savoir-faire oublié: Le travail du lait en Chine ancienne," *Zinbun*, no. 21 (1986): 31–65; Françoise Sabban, "Transition nutritionelle et histoire de la consummation laitière en Chine," no. 120 (2010): 1–6. H. T. Huang, *Science and Civilisation in China* 6:253.

17. Miranda Brown, "Mr. Song's Cheeses: Southern China, 1368–1644," *Gastronomica* 19, no. 2 (Summer 2019): 29–42.

18. Sun Zhongtang 孙忠堂, "Wai tai mi yao fang yaowu wenxian yanjiu 外台秘要方药物文献研究," in *Wai tai mi yao fang* 外台秘要方 [Arcane essential formulas from the Imperial Library], edited by Wang Tao 王涛 (Beijing: Huaxia Press, 1993), 1054.

19. These various uses of milk appear scattered throughout Wang, *Wai tai mi yao fang*, 127, 196, 198, 200, 202–204, 206, 231, 752.

20. Wang, *Wai tai mi yao fang*, 196, 198, 200, 202–204, 206, 231, 752.

21. On women in Malaysia, personal communication, Geoffrey Pakiam, July 2015; on Ayurvedic uses and southern Europe, Andrea Wiley, *Cultures of Milk: The Biology and Meaning of Dairy Products in the United States and India* (Cambridge, MA: Harvard University Press, 2014), 101, 104; on GI crises in China, Miranda Brown and Yong Yang, "Tummy Troubles: When Chinese Healers Administered Milk," *Journal of Chinese History* (January 2026): 1–19.

22. Wang Fengzhan 王凤展 and Yu Xinzhong 余新忠, "Niunai de jindaixing: Yi yingyang he weisheng wei zhongxin de sikao 牛奶的近代性:以营养和卫生为中心的思考" [The modernity of milk: Reflections centered on nutrition and health], *Zhongguo shehui lishi pinglun* 中国社会历史评论 [Chinese social history review] 16, no. 2 (2015): 3.

23. Wang and Yu, "Niunai de jindaixing," 3.

24. DuPuis, *Nature's Perfect Food*, chaps. 2 and 3.

25. R. D. Milner, *The Use of Milk as Food* (Washington, DC: Government Printing Office, 1909): 7.

26. Charles E. Jennings, "The Intravenous Injection of Milk," *British Medical Journal* 1275 (June 6, 1885): 1147–1149.

27. Alma Igra, "Over Spilt Milk: British Scientific Humanitarianism and the Quest for International Standards," *Isis* 115, no. 2 (June 2024): 335–353; Richard Wagner, *Clemens von Pirquet: His Life and Work* (Baltimore, MD: Johns Hopkins University Press, 1968).

28. Huang, *Science and Civilisation in China*, 6:256; Wang and Yu, "Niunai de jindaixing"; Françoise Sabban, Sabban, "The Taste for Milk in Modern China (1865–1937)," in *Food Consumption in Global Perspective*, edited by Anne Murcott and Jakob A. Klein, 182–209 (Basingstoke: Palgrave Macmillan, 2014).

29. Samuel J. Crumbine and James A. Tobey, *The Most Nearly Perfect Food: The Story of Milk* (Baltimore, MD: Williams & Wilkins Company, 1929), 77–78.

30. National Dairy Council pamphlet, as quoted in DuPuis, *Nature's Perfect Food*, 117.

31. Gan Rudeng 甘汝登, "Niuru yu jiankang zhi guanxi 牛乳與健康之關係" [The relationship between cow's milk and health], *Nongshi yuekan* 農事月刊 [Farming monthly] 3, no. 6 (1924): 17.

32. Gan, "Niuru yu jiankang zhi guanxi," 20.

33. My translation of the quotation in Wang and Yu, "Niunai de jindaixing," 21.

34. Wang and Yu, "Niunai de jindaixing," 22.

35. Bi Hui 皕誨, "Niuru yangsheng zhi fa 牛乳養生之法" [Ways of nourishing life through cow's milk], *Jinbu* 進步 [Progress] 8, no. 1 (1915): 82–83.

36. Wang and Yu, "Niunai de jindaixing," 20.

37. Dou Weilian 竇維廉 [William H. Adolph], "Zhongguo shiwu zhi chengfen ji qi yingyang jiazhi 中國食物之成分及其營養價值" [The composition and nutritional value of Chinese foodstuffs], *Kexue* 科學 [Science] 11, no. 8 (1926): 1032.

38. Jia-Chen Fu, *The Other Milk: Reinventing Soy in Republican China* (Seattle: University of Washington Press, 2018); Jia-Chen Fu, "Confronting the Cow: Soybean Milk and the Fashioning of a Chinese Dairy Alternative," in *Moral Foods: The Construction of Nutrition and Health in Modern Asia*, edited by Angela Ki Che Leung and Melissa L. Caldwell, 47–65 (Honolulu: University of Hawai'i Press, 2019).

39. Li Zhongping 李忠萍, "Cong jindai niuru guanggao kan Zhongguo de xiandaixing—yi 1927–1937 nian *Shen Bao* wei zhongxin de kaocha 从近代牛乳广告看中国的现代性— 以1927–1937年《申报》为中心的考察" [Modernity of China viewed from milk advertisements in Shanghai Post (*Shen Bao*) issued from 1927 to 1937], *Anhui daxue xuebao (Zhexue shehui kexue ban)* 安徽大学学报（哲学社会科学版）[Anhui University journal (Philosophy and social sciences edition)]. no. 3 (2010): 106–113. See also Susan Glosser, "Milk for Health, Milk for Profit: Shanghai's Chinese Dairy Industry under Japanese Occupation," in *Inventing Nanjing Road: Commercial Culture in Shanghai, 1900–1945*, edited by Sherman Cochran, 207–233 (Ithaca, NY: East Asia Program, Cornell University, 1999).

40. Wiley, *Cultures of Milk*, 109. Sabban, "Un savoir-faire oublié."

41. Wang and Yu, "Niunai de jindaixing," 21.

42. DuPuis, *Nature's Perfect Food*, chaps. 2 and 3.

43. Wang and Yu, "Niunai de jindaixing," 7.

44. Sabban, "The Taste for Milk in Modern China," 186–187.

45. *Final Report of the Mixed Committee of the League of Nations on the Relation of Nutrition to Health, Agriculture and Economic Policy* (Geneva: League of Nations, 1937), 89.

46. League of Nations, *The Problem of Nutrition*, vol. 3, *Nutrition in Various Countries* (Geneva: League of Nations, 1936).

47. Zhonghua yixue hui gonggong weisheng weiyuanhui tezu yingyang weiyuanhui 中华医学会公共卫生委员会特组营养委员会 [Special Working Group on Nutrition, Chinese Medical Society Public Health Committee], "Zhongguo minzhong zui di xiandu zhi yingyang xuyao 中国民众最低限度之营养需要" [Minimum nutritional requirements for the Chinese people], *Yingyang xuebao* 营养学报 [Acta nutrimenta sinica], originally published 1938, reprinted in 23, no. 3 (2001): 198. See also Jia-Chen Fu, "Would Mr. Science Eat the Chinese Diet?" *EASTS* 16, no. 3 (2022): 367–386.

48. Food and Agriculture Organization of the United Nations, *World Food Survey* (Washington, DC: 1946), 8.

49. Food and Agriculture Organization, *World Food Survey*, 17.

50. Valenze, *Milk*, 274.

51. An Act to Continue the Special Milk Program for Children," Pub. L. No. 85–478, July 1, 1958. Internet Archive URL: https://web.archive.org/web/20140714031651/http://www.gpo.gov/fdsys/pkg/STATUTE-72/pdf/STATUTE-72-Pg276-2.pdf.

52. Sophia Murphy and Kathy McAfee, *U.S. Food Aid: Time to Get it Right* (Minneapolis, MN: Institute for Agriculture and Trade Policy, 2005), 3–4.

53. Chen Yu-Jen 陳玉箴, "Yifu yu jingzheng: Zhan hou chuqi Mei yuan xia de Taiwan ruye (1945–1965) 依附與競爭：戰後初期美援下的台灣乳業(1945–1965)" [Dependence and

competition: The dairy industry in Taiwan during the post-war and US-aid period (1945–1965)], *Zhongguo yinshi wenhua* 中國飲食文化 [Chinese dietary culture] 13, no. 1 (2017): 35–73.

54. Chen, "Yifu yu jingzheng," 61. See also Smil and Kobayashi, *Japan's Dietary Transition and Its Impacts* (Cambridge, MA: MIT Press, 2012), 56.

55. Chen Yixing 陳怡行, ed., *Shi guang, shi guang—Taiwan yinshi dang'an gushi* 食光·時光—台灣飲食檔案故事 [Eras and the glory of food—Stories from Taiwan's food archives] (Hsin bei: National Development Committee Archive Management Bureau/Ministry of Culture/Hakka Committee Hakka Cultural Development Center, 2021), 68.

56. Thomas David DuBois, "China's Dairy Century: Making, Drinking and Dreaming of Milk," in *Animals and Human Society in Asia: Historical, Cultural and Ethical Perspectives*, edited by Rotem Kowner et al., 179–212 (Cham, Switzerland: Palgrave Macmillan, 2019).

57. DuBois, "China's Dairy Century," 197.

58. Han Guang 韩光, "Yijiuqiba nian quanguo rupin shengchan chuangzao le jianguo yilai zui hao shuiping 一九七八年全国乳品生产创造了建国以来最好水平" [In 1978 national dairy-product production reached its best level since the establishment of the PRC], *Rupin gongye* 乳品工业 [China dairy industry] no. 3 (June 30, 1979): 1.

59. Don P. Blayney, "The Changing Landscape of U.S. Milk Production," *United States Department of Agriculture Statistical Bulletin No.978* (June 2002), 2. Internet Archive URL: https://web.archive.org/web/20130308102307/http://www.ers.usda.gov/media/488847/sb978_1_.pdf.

60. Wiley, "Milk for 'Growth,'" 14.

61. Louisa Lim, "China's Growing Thirst for Milk Hits Global Market," NPR News, April 21, 2008). Internet Archive URL: https://web.archive.org/web/20080420171250/http://www.npr.org/templates/story/story.php?storyId=89766337; Wiley, "Milk for 'Growth,'" 18–20.

62. Tania Branigan, "Chinese Figures Show Fivefold Rise in Babies Sick from Contaminated Milk," *The Guardian*, December 2, 2008. Internet Archive URL: https://web.archive.org/web/20130905090333/http://www.theguardian.com/world/2008/dec/02/china.

63. Shi Zhifang 石智芳, Wang Huantang 王焕棠, and Wu Pufeng 吴普峰, "Niunai yu ren de jiankang 牛奶与人的健康" [Milk and human health], *Shanxi shipin gongye* 山西食品工业 [Shanxi food products industry] no. 4 (December 1999): 41.

64. Wiley, "Milk for 'Growth,'" 22.

65. Shi et al., "Niunai yu ren de jiankang," 41.

66. "Wen Jiabao Chongqing kaocha: rang meige Zhongguoren mei tian he shang yi jin nai 温家宝重庆考察：让每个中国人每天喝上一斤奶" [Wen Jiabao's Chongqing inspection: Let every Chinese person drink a *jin* of milk every day], *Xinhua News Agency* 新华, April 25, 2006. Internet Archive URL: https://web.archive.org/web/20250717172831/https://finance.sina.cn/sa/2006-04-25/detail-ikknscsi3312344.d.html.

67. *Zhongguo yingyang xuehui* 中国营养学会 [Chinese Nutrition Society], *Zhongguo jumin shanshi zhinan de fazhan zhuangkuang* 中国居民膳食指南的发展状况 [A history of the development of the Chinese dietary guidelines]. Internet Archive URL: https://web.archive.org/web/20250207042026/http://dg.en.cnsoc.org/article/lsqy.html.

68. Qiqing Lin and Colum Murphy, "Spilled Milk: How China Failed Its Dairy Farmers," *Sixth Tone: Fresh Voices from Today's China*, December 19, 2017. Internet Archive

URL: https://web.archive.org/web/20250625010008/https://www.sixthtone.com/news/1001417.

69. Curry, "Archaeology," 20–22; Huang, "Hypolactasia and the Chinese Diet"; Kenneth Kiple, "Lactose Intolerance," in *The Cambridge World History of Food*, edited by Kenneth Kiple, section IV.E.6 (Cambridge: Cambridge University Press, 2000).

70. Theodore M. Bayless and Norton S. Rosensweig, "A Racial Difference in Incidence of Lactase Deficiency: A Survey of Milk Intolerance and Lactase Deficiency in Healthy Adult Males," *JAMA* 197, no. 12 (September 19, 1966): 139.

71. Bayless and Rosensweig, "A Racial Difference," 140.

72. George Alexander, "Something for Every Body—But Is It Good?" *Los Angeles Times*, April 14, 1974, 3; "They Just Might Outgrow Need for Milk," *Washington Post*, October 7, 1971, C3; Joshua Lederberg, "A Hereditary Tummy-Ache?" *Washington Post*, November 20, 1966, E7.

73. Bayless and Rosensweig, "A Racial Difference," 140–141.

74. Sidney C. Alford, "Lactose Intolerance in Asians," *Nature* 221 (February 8, 1969): 562–563; T. D. Bolin and A. E. Davis, "Asian Lactose Intolerance and Its Relation to Intake of Lactose," *Nature* 222 (April 26, 1969): 382–383; Myung H. Chung and Douglas B. McGill, "Lactase Deficiency in Orientals," *Gastroenterology* 54, no. 2 (1968): 225–226; S. S. Huang and T. M. Bayless, "Milk and Lactose Intolerance in Healthy Orientals," *Science* 160, no. 3823 (April 5, 1968): 83–84; A. E. Davis and T. Bolin, "Lactose Intolerance in Asians," *Nature* 216 (December 23, 1967): 1244–1245.

75. K. L. Chua and C. S. Seah, "Lactose Intolerance: Hereditary or Acquired? Effect of Prolonged Milk Feeding," *Singapore Medical Journal* 14, no. 1 (March 1973): 29–33.

76. Frederick J. Simoons, "Primary Adult Lactose Intolerance and the Milking Habit: A Problem in Biologic and Cultural Interrelations, II. A Culture Historical Hypothesis," *American Journal of Digestive Diseases* 15, no. 8 (August 1970): 699.

77. David M. Paige and Theodore M. Bayless, eds., *Lactose Digestion: Clinical and Nutritional Implications* (Baltimore, MD: Johns Hopkins University Press, 1981), 7.

78. Lederberg, "A Hereditary Tummy-Ache?"

79. Bayless and Rosensweig, "A Racial Difference," 142.

80. David M. Paige, Theodore M. Bayless, and George G. Graham, "Milk Programs: Helpful or Harmful to Negro Children?" *American Journal of Public Health* 62, no. 11 (November 1972): 1486–1488.

81. Robert B. Bradfield, D. B. Jelliffe, and Aaron Ifekwunigwe, "Milk Intolerance and Malnutrition," *The Lancet* 2, no. 7929 (August 16, 1975): 325.

82. Quoted in a letter to the editor by R.L. Strock, *Baltimore Sun*, November 12, 1972.

83. Thomas J. Marchione, "Foods Provided through U.S. Government Emergency Food Aid Programs: Policies and Customs Governing their Formulation, Selection and Distribution," *Journal of Nutrition* 132, no. 7 (July 1, 2002): 2107S.

84. *Rutangmei ji di rutang ruzhipin zhuanti yanjiuhui wenji* 乳糖酶及低乳糖乳制品专题研究会文集 [Proceedings of the conference on lactase and low-lactose dairy products] (Beijing: Chinese Nutrition Society, 2001).

85. "Zhongguo peiyu de shouli zhuanjiyin di rutang nainiu you wang 2014 nian chan nai 中国培育的首例转基因低乳糖奶牛有望2014年产奶" [The first genetically engineered

low-lactose dairy cow, Chinese-bred, may produce milk in 2014], *Hubei xumu shouyi* 湖北畜牧兽医 [Hubei journal of animal and veterinary sciences] 5 (May 5, 2013): 56.

86. Xiaohu Su et al., "Production of Microhomologous-Mediated Site-Specific Integrated LacS Gene Cow Using TALENs," *Theriogenology* 119 (October 1, 2018): 282–288. Thanks to Anne Mendelson for pointing me to this source.

87. "They Just Might Outgrow Need for Milk," C3.

88. Anthony K. Campbell, Jonathan P. Waud, and Stephanie B. Matthews, "The Molecular Basis of Lactose Intolerance," *Science Progress* 92, nos. 3–4 (2009): 241–287.

89. Albert D. Newcomer, "Immediate Symptomatic and Long-Term Nutritional Consequences of Hypolactasia," in *Lactose Digestion: Clinical and Nutritional Implications*, edited by David M. Paige and Theodore M. Bayless, 124–133 (Baltimore, MD: Johns Hopkins University Press, 1981).

90. Vasile Stănescu, "'White Power Milk': Milk, Dietary Racism, and the 'Alt-Right,'" *Animal Studies Journal* 7, no. 2 (2018): 114–115.

91. Amy Harmon, "Why White Supremacists Are Chugging Milk (and Why Geneticists Are Alarmed)," *New York Times*, October 17, 2018. Internet Archive URL: https://web.archive.org/web/20181017091349/https://www.nytimes.com/2018/10/17/us/white-supremacists-science-dna.html.

92. Iselin Gambert and Tobias Linné, "How the Alt-Right Uses Milk to Promote White Supremacy," *The Conversation*, April 26, 2018. Internet Archive URL: https://web.archive.org/web/20180426155747/https://theconversation.com/how-the-alt-right-uses-milk-to-promote-white-supremacy-94854.

93. C. Justin Cook, "The Role of Lactase Persistence in Precolonial Development," *Journal of Economic Growth* 19, no. 4 (December 2014): 398.

94. "No Use Crying: The Ability to Digest Milk May Explain How Europe Got Rich," *The Economist*, March 28, 2015.

95. Richard Cammack, "Lactose Intolerance," in *Oxford Dictionary of Biochemistry and Molecular Biology*, edited by Richard Cammack et al. (Oxford: Oxford University Press, 2006), 375.

Chapter 5. Temperance as Sickness

1. Li Shang-jen 李尚仁, "Jiankang de daode jingji—De Zhen lun Zhongguoren de shenghuo xiguan he weisheng 健康的道德經濟— 德貞論中國人的生活習慣和衛生" [The moral economy of health—John Dudgeon on Chinese people's habits and hygiene], in *Diguo yu xiandai yixue* 帝國與現代醫學 [Imperialism and modern medicine], edited by Li Shang-jen (Taipei: Lianjing, 2008), 234–235.

2. J. G. Kerr, "Drinking Habits of Chinese," *Chinese Recorder and Missionary Journal*, September 1, 1870, 85.

3. W. H. Park, "Alcohol in China. Interesting Reminiscences. Drunkenness Common. A Doctor's Experiences," *Shanghai Times*, October 23, 1916.

4. Shailer Mathews, "Shall We Make the Chinese Drunkards?" *The Independent* 104, no. 374 (November 6, 1920): 187.

5. Patrick E. McGovern, *Uncorking the Past: The Quest for Wine, Beer and Other Alcoholic Beverages* (Berkeley: University of California Press, 2009).

6. Mu-Chou Poo, "The Use and Abuse of Wine in Ancient China," *Journal of the Economic and Social History of the Orient* 42, no. 2 (1999): 127.

7. Roel Sterckx, "Alcohol and Historiography in Early China," *Global Food History* 1, no. 1 (2015): 13–32.

8. Tak Kam Chan, "From Conservatism to Romanticism: Wine and Prose-Writing from Pre-Qin to Jin," in *Scribes of Gastronomy: Representations of Food and Drink in Imperial Chinese Literature*, edited by Isaac Yue and Siufu Tang, 15–26 (Hong Kong: Hong Kong University Press, 2013).

9. Sterckx, "Alcohol and Historiography in Early China"; Poo, "The Use and Abuse of Wine."

10. Nicholas Morrow Williams, "The Morality of Drunkenness in Chinese Literature of the Third Century CE," in *Scribes of Gastronomy: Representations of Food and Drink in Imperial Chinese Literature*, edited by Isaac Yue and Siufu Tang, 27–45 (Hong Kong: Hong Kong University Press, 2013).

11. Charles Kwong, "Making Poetry with Alcohol: Wine Consumption in Tao Qian, Li Bai and Su Shi," in *Scribes of Gastronomy: Representations of Food and Drink in Imperial Chinese Literature*, edited by Isaac Yue and Siufu Tang, 45–68 (Hong Kong: Hong Kong University Press, 2013).

12. Scholars usually leave the term *zang*, as in *zang* organs, untranslated. For a brief explanation of *zang* organs, please see chapter 2, note 8.

13. Edwin van Bibber-Orr, "Alcoholism and Song Literati," in *Behaving Badly in Early and Medieval China*, edited by N. Harry Rothschild and Leslie V. Wallace, 135–153 (Honolulu: University of Hawai'i Press, 2017).

14. Kwong, "Making Poetry with Alcohol," 53.

15. Translation by van Bibber-Orr, "Alcoholism and Song Literati," 142.

16. American Psychiatric Association, *Diagnostic and Statistical Manual of Mental Disorders*, Fifth Edition (DSM-5-TR) (Washington, DC: American Psychiatric Association Publishing, 2022), 554.

17. See Hilary A. Smith, *Forgotten Disease: Illnesses Transformed in Chinese Medicine* (Stanford, CA: Stanford University Press, 2017).

18. James A. Benn, "Buddhism, Alcohol and Tea in Medieval China," in *Of Tripod and Palate: Food, Politics, and Religion in Traditional China*, edited by Roel Sterckx, 213–236 (New York: Palgrave Macmillan, 2005).

19. Liao Xiuping 廖秀萍 et al., "Mantan Huaxia jiu wenhua yu Zhongyi yong jiu zhi dao 漫谈华夏酒文化与中医用酒之道" [An informal discussion of Chinese alcohol culture and the Chinese medicine way of using alcohol], *Zhongguo Zhongyiyao xiandai yuancheng jiaoyu* 中国中医药现代远程教育 [Chinese pharmacology modern distance learning] 21, no. 23 (December 2023): 127.

20. Yan Liu, *Healing with Poisons: Potent Medicines in Medieval China* (Seattle: University of Washington Press, 2021).

21. Xia Xun-li, "History of Chinese Medicinal Wine," *Chinese Journal of Integrative Medicine* 19, no. 7 (July 2013): 549–555; Sun Longchuan 孙龙川, "Jiu yu Zhongyi zhiliao zhong de yiyi 酒于中医治疗中的意义" [The significance of alcohol in Chinese medicine therapy], *Jiangxi Zhongyiyao* 江西中医药 [Jiangxi Chinese medicine] no. 4 (May 1, 1986): 49–50.

22. Bob Flaws, *Chinese Medicinal Wines and Elixirs* (Boulder, CO: Blue Poppy Press, 1994).

23. Wang Xuetai 王学泰, *Zhongguo yinshi wenhua shi* 中国饮食文化史 [A history of Chinese dietary culture] (Beijing: Zhongguo qingnian chubanshe, 2012), 189. See also Kwong, "Making Poetry with Alcohol," 61–62.

24. Norman Smith, *Intoxicating Manchuria: Alcohol, Opium, and Culture in China's Northeast* (Vancouver: University of British Columbia Press, 2012). Although pre-twentieth-century observers could not have known this, a genetic factor may also have contributed to this observed difference. Scientists have noted that the genotype that produces so-called aldehyde dehydrogenase deficiency is more common in people from south China and Taiwan than in people from north China. Mimy Y. Eng, Susan E. Luczak, and Tamara L. Wall, "*ALDH2, AHD1B*, and *ADH1C* Genotypes in Asians: A Literature Review," *Alcohol Research and Health* 30, no. 1 (2007): 22–27.

25. Volker Scheid, "From Civilizing Foods for a Nourishing Life to a Global Traditional Chinese Medicine Dietetics," in *Moral Foods: The Construction of Nutrition and Health in Modern Asia*, edited by Angela Ki Che Leung and Melissa L. Caldwell, 241–261 (Honolulu: University of Hawai'i Press, 2019).

26. Xu Hongbin 許宏彬, "Cong apian junzi dao jiaozheng yangben: apian xishizhe, gengsheng yuan yu Du Congming 從阿片君子到矯正樣本：阿片吸食者、更生院與杜聰明" [From opium fiend to specimen for rectification: Opium users, the Resurrection Hospital, and Du Congming], in *Diguo yu xiandai yixue* 帝國與現代醫學 [Imperialism and modern medicine], edited by Li Shang-jen, 101–146 (Taipei: Lianjing, 2008).

27. Zhiguo Yang, "'This Beer Tastes Really Good': Nationalism, Consumer Culture and Development of the Beer Industry in Qingdao, 1903–1993," *Chinese Historical Review* 14, no. 1 (2013): 29–58.

28. Yang, "'This Beer Tastes Really Good,'" 50.

29. Gerard Sasges, *Imperial Intoxication: Alcohol and the Making of Colonial Indochina* (Honolulu: University of Hawai'i Press, 2017); Erica J. Peters, *Appetites and Aspirations in Vietnam: Food and Drink in the Long Nineteenth Century* (Lanham, MD: AltaMira Press, 2012).

30. Mathews, "Shall We Make the Chinese Drunkards?"

31. Park, "Alcohol in China," 7.

32. Mathews, "Shall We Make the Chinese Drunkards?"

33. "Exporting Drink to China," *Millard's Review of the Far East*, July 20, 1918, 311; "NATIONS AND DRINK: Protest against Alcoholising China to Peace Conference," *Shanghai Times*, July 3, 1919); Tom Doyle, "Is China as 'Dry' as She Is Painted?" *China Weekly Review*, June, 26, 1926.

34. Doyle, "Is China as 'Dry' as She Is Painted?"

35. Doyle, "Is China as 'Dry' as She Is Painted?"

36. Kenneth Pomeranz, "Political Economy and Ecology on the Eve of Industrialization: Europe, China, and the Global Conjuncture," *American Historical Review* 107, no. 2 (2002): 425–446.

37. Amy Mittelman, *Brewing Battles: A History of American Beer* (New York: Algora Publishing, 2008), 17.

38. Daniel Okrent, *Last Call: The Rise and Fall of Prohibition* (New York: Scribner, 2010) 39.

39. Smith, *Intoxicating Manchuria*, 68.

40. Jeffrey M. Pilcher, "'Tastes Like Horse Piss': Asian Encounters with European Beer," *Gastronomica* 16, no. 1 (Spring 2016): 37.

41. Yang, "'This Beer Tastes Really Good,'" 38–39.

42. Smith, *Intoxicating Manchuria*, 17.

43. Smith, *Intoxicating Manchuria*, 17.

44. See, for example, Chen Xiang 陳香, *Jiu Ling* 酒令 [Drinking games] (Taipei: Guo jia chu ban she, 1983); Liu Chutang 刘初棠, *Zhongguo gudai jiu ling* 中国古代酒令 [Chinese ancient drinking games] (Shanghai: Shanghai renmin chubanshe, 1993).

45. Smith, *Intoxicating Manchuria*, 17.

46. Rune Andréasson and A. Wayne Jones, "The Life and Work of Erik M. P. Widmark," *American Journal of Forensic Medicine and Pathology* 17, no. 3 (September 1996): 177–190.

47. J. G. Wigmore and R. M. Langille, "Six Generations of Breath Alcohol Testing Instruments: Changes in the Detection of Breath Alcohol Since 1930. An Historical Overview," *Canadian Society of Forensic Science Journal* 42, no, 4 (2009): 276–283.

48. Helge Kragh, "From Disulfiram to Antabuse: The Invention of a Drug," *Bulletin for the History of Chemistry* 33, no. 2 (2008): 82–87.

49. O. Martensen-Larsen, "Treatment of Alcoholism with a Sensitising Drug," *The Lancet* 252, no. 6539 (December 25, 1948): 1004–1005.

50. Erik Jacobsen, "The Metabolism of Ethyl Alcohol," *Nature* 169, no. 4303 (April 19, 1952): 645–647.

51. D. Fenna et al., "Ethanol Metabolism in Various Racial Groups," *Canadian Medical Association Journal* 105, no. 5 (September 4, 1971): 472.

52. Fenna et al., "Ethanol Metabolism in Various Racial Groups," 475.

53. An often-cited study that found no difference is L. J. Bennion and T. K. Li, "Alcohol Metabolism in American Indians and Whites: Lack of Racial Differences in Metabolic Rate and Liver Alcohol Dehydrogenase," *New England Journal of Medicine* 294, no. 1 (January 1976): 9–13. For a study that argues that Native subjects' metabolisms are faster than those of White subjects, see Arthur R. Zeiner et al., "Racial Differences in Psychophysiological Responses to Ethanol and Placebo," in *Currents in Alcoholism*, vol. 1, edited by Frank A. Seixas, 271–286 (New York: Grune & Stratton, 1977).

54. Sarah W. Tracy, *Alcoholism in America* (Baltimore, MD: Johns Hopkins University Press, 2005).

55. Weston La Barre, "Some Observations on Character Structure in the Orient. II. The Chinese, Part Two," *Psychiatry* 9, no. 4 (November 1, 1946): 376.

56. Milton L. Barnett, "Alcoholism in the Cantonese of New York City: An Anthropological Study," in *Etiology of Chronic Alcoholism*, edited by Oskar Diethelm, 179–227 (Springfield, IL: Charles C. Thomas, 1955).

57. Tsung-yi Lin, "A Study of the Incidence of Mental Disorder in Chinese and Other Cultures," *Psychiatry* 16, no. 4 (November 1, 1953): 334, 23.

58. Arthur L. Klatsky et al., "Racial Patterns of Alcoholic Beverage Use," *Alcoholism: Clinical and Experimental Research* 7, no. 4 (Fall 1983): 372–377; Kua Ee Heok, "Drinking in Chinese Culture: Old Stereotypes Re-Examined," *British Journal of Addiction* 82, no. 3 (1987): 224–225.

59. Morris E. Chafetz, "Consumption of Alcohol in the Far and Middle East," *New England Journal of Medicine* 271, no. 6 (August 6, 1964): 297–301.

60. Richard P. Wang, "A Study of Alcoholism in Chinatown," *International Journal of Social Psychiatry* 14, no. 4 (September 1968): 260–267.

61. Peter H. Wolff, "Ethnic Differences in Alcohol Sensitivity," *Science* New Series 175, no. 4020 (January 28, 1972): 449.

62. Wolff, "Ethnic Differences in Alcohol Sensitivity."

63. Daniel E. Moerman, *Meaning, Medicine, and the "Placebo Effect,"* (Cambridge: Cambridge University Press, 2002).

64. S. Harada, D. P. Agarwal, and H. W. Goedde, "Aldehyde Dehydrogenase Deficiency as Cause of Facial Flushing Reaction to Alcohol in Japanese." *The Lancet* 2, no. 8253 (October 31, 1981): 982.

65. Harada et al., "Aldehyde Dehydrogenase Deficiency."

66. S. Harada et al., "Possible Protective Roles against Alcoholism for Aldehyde Dehydrogenase Isozyme Deficiency in Japan," *The Lancet* 2, no. 8302 (October 9, 1982): 827.

67. Dharam P. Agarwal and H. Werner Goedde, "Human Aldehyde Dehydrogenases: Their Role in Alcoholism," *Alcohol* 6, no. 6 (1989): 517.

68. See, for example, Helen M. Chao, "Alcohol and the Mystique of Flushing," *Alcoholism: Clinical and Experimental Research* 19, no. 1 (1995): 104–109; Guang-Chou Tu and Yedy Israel, "Alcohol Consumption by Orientals in North America Is Predicted Largely by a Single Gene," *Behavior Genetics* 25, no. 1 (1995): 59–65; Kazufumi Dohmen et al., "Ethnic Differences in Gastric Sigma-Alcohol Dehydrogenase Activity and Ethanol First-Pass Metabolism," *Alcoholism: Clinical and Experimental Research* 20, no. 9 (December 1996): 1569–1576.

69. Che-Hong Chen et al., "Targeting Aldehyde Dehydrogenase 2: New Therapeutic Opportunities," *Physiological Reviews* 94, no. 1 (January 2014): 1–34.

70. Stanford Taiwan ALDH2 Consortium, "Stanford-Taiwan ALDH2 Deficiency Research (STAR) Consortium." Internet Archive URL: https://web.archive.org/web/20230517090843/https://envmed.kmu.edu.tw/images/Stanford-Taiwan_ALDH2_Deficiency_Research_Consortium_STAR_Mission__Scope_Statement_1.pdf.

71. Meeri Kim, "'Asian Glow' from Alcohol Isn't Just a Discomfort. It's a Severe Warning," *Washington Post*, August 15, 2023, https://www.washingtonpost.com/wellness/2023/08/15/asian-glow-alcohol-gene-mutation-cancer-risk/.

72. Yuki Matsumura et al., "Gene Therapy Correction of Aldehyde Dehydrogenase 2 Deficiency," *Molecular Therapy Methods & Clinical Development* 15 (August 26, 2019): 72–82.

73. Krystle Vermes, "Redwood City Startup, Aldea Pharmaceuticals, Reels in $24 Million," *Biospace*, August 25, 2014. Internet Archive URL: https://web.archive.org/web/20250717175854/https://www.biospace.com/redwood-city-startup-aldea-pharmaceuticals-reels-in-24-million.

74. The rival is New Horizons. See New Horizons Pharma homepage. Internet Archive URL: https://web.archive.org/web/20251217095342/https://www.newhorizonspharmalab.com/.

75. "The Science behind Pre-Alcohol," ZBiotics. Internet Archive URL: https://web.archive.org/web/20260201123150/https://zbiotics.com/pages/how-it-works.

76. "How the Anytime You Drink Vitamin Works," H-Proof. Internet Archive URL: https://web.archive.org/web/20240222230610/https://h-proof.com/pages/how-it-works.

77. "Learn the Science of Refreshment," DrinkAid. Internet Archive URL: https://web.archive.org/web/20260116055612/https://drinkaid.co/pages/science.

78. I have been unable to ascertain definitively when the company closed, but the Crunchbase site suggests it may have been in 2017; see https://web.archive.org/web/20140811055604/http://www.crunchbase.com/organization/aldea-pharmaceuticals.

79. Robert Aronowitz, *Risky Medicine: Our Quest to Cure Fear and Uncertainty* (Chicago: University of Chicago Press, 2015).

80. Jakob Manthey et al., "Global Alcohol Exposure between 1990 and 2017 and Forecasts until 2030: A Modelling Study," *The Lancet* 393, no. 10190 (June 22, 2019): 2501.

81. Yi-lang Tang et al., "Alcohol and Alcohol-Related Harm in China: Policy Changes Needed," *Bulletin of the World Health Organization* 91, no. 4 (April 2013): 270–276. The "other western countries" quotation is from Hui G. Cheng et al., "Prevalence of Alcohol Use Disorders in Mainland China: A Systematic Review," *Addiction* 110, no. 5 (May 2015): 761–774.

82. Jeffrey Pilcher, Wang Yu, and Yuebing Jackson Guo, "'Beer with Chinese Characteristics': Marketing Beer under Mao," *RAE Revista de Administração de Empresas* [Journal of business management] 58, no. 3 (May/Jun 2018): 313.

83. Xu Guo and Yong-guang Huang, "The Development of Alcohol Policy in Contemporary China," *Journal of Food and Drug Analysis* 23, no. 1 (March 2015): 19–29; Aqian Hu et al., "The Transition of Alcohol Control in China 1990–2019: Impacts and Recommendations," *International Journal of Drug Policy* 105, no. 103698 (July 2022): 1–9. On Taiwan, see "Taiwan yanjiu shiye: cong zhuanmai dao ziyouhua jingying 台灣菸酒事業：從專賣到自由化經營" [Taiwan tobacco and alcohol institutions: From monopoly to liberalized management], *Dang'an lehuo qingbao* 檔案樂活情報 [Archives LOHAS] no. 193 (July 17, 2023). Internet Archive URL: https://web.archive.org/web/20241014163937/https://www.archives.gov.tw/tw/arctw/69-1738.html.

84. Li Jinhua 李金华 et al., "Yunnan Yizu jiujing yilai zonghezheng yu ALDH2 jiyin duotaixing de guanlian yanjiu 云南彝族酒精依赖综合征与 ALDH2 基因多态性的关联研究" [Research into the connection between alcohol-dependence syndrome among the Yi people of Yunnan and ALDH2-gene polymorphism], *Zhongguo yaowu yilaixing zazhi* 中国药物依赖性杂志 [Chinese journal of drug dependence] 22, no. 4 (2013): 263–266, 291.

85. Shaomin Yan and Guang Wu, "Could ALDH2*2 Be the Reason for Low Incidence and Mortality of Ovarian Cancer for East Asia [sic] Women?" *Oncotarget* 9, no.15 (February 23, 2018): 12503–12512.

86. Huang Jianshi 黄建始, "Haoyin xiaozhuo jie shang shenxin 豪饮小酌皆伤身心" [Unrestrained drinking and having just a few drinks both harm your body and mind], *Jiankang guanli* 健康管理 [Health management] 2, no. 6 (June 2011): 10–14.

87. Y.-P. Lin and T.-J. Cheng, "Why Can't Chinese Han Drink Alcohol? Hepatitis B Virus Infection and the Evolution of Acetaldehyde Dehydrogenase Deficiency," *Medical Hypotheses* 59, no. 2 (2002): 204–207.

88. Katherine J. Wu, "What If There's a Secret Benefit to Getting Asian Glow?" *The Atlantic*, (October 25, 2023). Internet Archive URL: https://web.archive.org/web

/20260214044311/https://www.theatlantic.com/science/archive/2023/10/alcohol-flush-asian-genetic-mutation-cause/675759/.

89. Lin and Cheng, "Why Can't Chinese Han Drink Alcohol?"; Wu, "What If There's a Secret Benefit to Getting Asian Glow?"

90. My translation of the passage as quoted in Wang Saishi 王赛时, *Zhongguo jiu shi* 中国酒史 [A History of Alcohol in China] (Ji'nan: Shandong University Press, 2010): 356.

Conclusion

1. Hsien Wu (Wu Xian 吳憲), "Nutritional Deficiencies in China and Southeast Asia," in *Proceedings of the Fourth International Congress on Tropical Medicine and Malaria,* vol. 2 (Washington, DC: May 10–18, 1948), 1222.

2. Shang-Jen Li, "Eating Well in China: British Medical Men on Diet and Personal Hygiene at Nineteenth-Century Chinese Treaty Ports," in *Health and Hygiene in Chinese East Asia: Policies and Publics in the Long Twentieth Century*, edited by Angela Ki Che Leung and Charlotte Furth, 109–131 (Durham, NC: Duke University Press, 2010).

3. Ian Hacking, "Why Race Still Matters," *Daedalus* 134, no. 1 (Winter 2005): 102–116.

4. Timothy Grose, "Beijing's Culinary Crusade: Erasing Uyghur Identity through Food," *ChinaFile*, May 13, 2024. Internet Archive URL: https://web.archive.org/web/20260131033819/https://www.chinafile.com/reporting-opinion/viewpoint/beijings-culinary-crusade-erasing-uyghur-identity-through-food.

5. Wen-Harn Pan et al., "Healthy Taiwanese Eating Approach (TEA) toward Total Wellbeing and Healthy Longevity," *Nutrients* 14, no. 2774 (2022): 1–13.

6. Fan Zhihong 范志红, "'Dongfang shanshi moshi' daodi hao zai na'er? Gai zenme chi?「东方膳食模式」到底好在哪儿？该怎么吃?" [What are the benefits of the 'Eastern diet model'? How should one eat it?]. *Beijing qingnian bao* 北京青年报 [Beijing youth news]. Internet Archive URL: https://web.archive.org/web/20250116090029/http://www.news.cn/food/20220602/f286ff5634804880a94e9f2a567c2201/c.html.

7. "Liu Xiqian: 'Zhongguo tese shuang danbai gongcheng' rang guo ren geng jiankang 刘锡潜：'中国特色双蛋白工程"让国人更健康'" [Liu Xiqian: "Dual protein with Chinese characteristics" will make Chinese people healthier], *Zhongguo jingji wang* 中国经济网 [China economy net], March 3, 2012. Internet Archive URL: https://web.archive.org/web/20211129034659/http://www.ce.cn/cysc/sp/info/201203/03/t20120303_21125031.shtml.

8. "Guowuyuan bangongting guanyu yinfa guomin yingyang jihua (2017–2030 nian) de tongzhi 国务院办公厅关于印发国民营养计划(2017–2030年)的通知" [Announcement from the State Council General Office regarding the printing and distribution of the Citizens' Nutrition Plan (2017–2030)], *Zhongguo zhengfu wang* 中国政府网 [China government net], July 13, 2017. Internet Archive URL: https://web.archive.org/web/20251208125215/https://www.gov.cn/zhengce/content/2017-07/13/content_5210134.htm.

9. "Shou jie guoji shuang danbai yu chuangxin danbai chanye fazhan dahui ji shuang danbai shiwu hangye biaozhun tuijin hui chenggong zhaokai 首届国际双蛋白与创新蛋白产业发展大会暨双蛋白食物行业标准推进会成功召开" [First international dual protein and innovative protein production conference and dual-protein industry standards advancement conference successfully convened], *Nongye nongcun bu shiwu yu yingyang fazhan yanjiusuo* 农业农村部食物与营养发展研究所 [Institute of Food and Nutrition Development, Ministry of Agriculture and Rural Affairs], December 22, 2023. Internet Archive URL:

https://web.archive.org/web/20251107000751/https://ifnd.caas.cn/xwzx/yw/08826219756d4d81a9d20d62c862b197.htm.

10. T. Colin Campbell and Thomas M. Campbell, *The China Study: The Most Comprehensive Study of Nutrition Ever Conducted and the Startling Implications for Diet, Weight Loss, and Long-Term Health* (Dallas, TX: BenBella Books, 2006).

11. Bob Flaws, *The Tao of Healthy Eating: Dietary Wisdom According to Traditional Chinese Medicine* (Boulder, CO: Blue Poppy Press, 1998); Joerg Kastner, *Chinese Nutrition Therapy: Dietetics in Traditional Chinese Medicine (TCM)* (Stuttgart: Thieme, 2004); John Shi, Chi-Tang Ho, and Fereidoon Shahidi, *Functional Foods of the East* (Boca Raton, FL: CRC Press, 2010).

12. Stacey Isaacs, *The Chinese Medicine Cookbook: Nourishing Recipes to Heal and Thrive* (Emeryville, CA: Althea Press, 2019).

13. "The Shorter Discourse to Māluṇkyaputta," in *Early Buddhist Discourses*, edited and translated by John J. Holder, 95–100 (Indianapolis, IN: Hackett, 2006).

Index

Printed and bound by CPI Group (UK) Ltd, Croydon, CR0 4YY
07/10/2026
14977462-0003